Simulating
Nearshore
Environments

COMPUTER METHODS IN THE GEOSCIENCES

Daniel F. Merriam, Series Editor

Volumes in the series published by Pergamon:

Geological Problem Solving with Lotus 1-2-3 for Exploration and Mining Geology: G.S. Koch Jr. (with program on diskette)
Exploration with a Computer: Geoscience Data Analysis Applications: W.R. Green
Contouring: A Guide to the Analysis and Display of Spatial Data: D.F. Watson (with program on diskette)
Management of Geological Data Bases: J Frizado

*Volumes published by Van Nostrand Reinhold Co. Inc.:

Computer Applications in Petroleum Geology: J.E. Robinson
Graphic Display of Two-and Three-Dimensional Markov Computer Models in Geology: C. Lin and J.W. Harbaugh
Image Processing of Geological Data: A.G. Fabbri
Contouring Geologic Surfaces with the Computer: T.A. Jones, D.E. Hamilton, and C.R. Johnson
Exploration-Geochemical Data Analysis with the IBM PC: G.S. Koch, Jr. (with programs on diskettes)
Geostatics and Petroleum Geology: M.E. Hohn
Simulating Clastic Sedimentation: D.M. Tetzlaff and and J.W. Harbaugh

*Orders to: Van Nostrand Reinhold Co. Inc, 7625, Empire Drive, Florence, KY 41042, USA.

Related Pergamon Publications

Books

GAAL & MERRIAM (Editors): Computer Applications in Resource Estimation: Prediction and Assessment for Metals and Petroleum

HANLEY & MERRIAM (Editors): Microcomputer Applications in Geology I and II

OPEN UNIVERSITY: Waves, Tides and Shallow Water Processes

Journals

Computers & Geosciences

Continental Shelf Research

Full details of all Pergamon publications/free specimen copy of any Pergamon journal available on request from your nearest Pergamon office.

Simulating Nearshore Environments

Paul A Martinez
Occidental Petroleum Company,
Bakersfield,
California,
USA

and

John W Harbaugh
Professor of Applied Earth Sciences,
Stanford University,
Stanford,
California,
USA

PERGAMON PRESS

OXFORD · NEW YORK · SEOUL · TOKYO

U.K.	Pergamon Press Ltd, Headington Hill Hall, Oxford OX3 0BW, England
U.S.A.	Pergamon Press, Inc, 660 White Plains Road, Tarrytown, New York 10591-5153, U.S.A.
KOREA	Pergamon Press Korea, KPO Box 315, Seoul 110-603, Korea
JAPAN	Pergamon Press Japan, Tsunashima Building Annex, 3-20-12 Yushima, Bunkyo-ku, Tokyo 113, Japan

First edition 1993

Library of Congress Cataloging in Publication Data
A catalogue record for this book is available from the Library of Congress.

British Library Cataloguing in Publication Data
A catalogue record for this book is available from the the British Library.

ISBN 0 08 037937 0

The cover illustration was generated with software from Dynamic Graphics Inc (Alameda, CA) on a Silicon Graphics (Mountain View, CA) IRIS workstation.

Printed in Great Britain by BPCC Wheatons Ltd., Exeter

Contents

Series Editor's Foreword

Simulation of geological processes is an important element in understanding geological situations. Simulations, or modeling, can utilize different parameters and manipulate them to learn of the responses to certain changes. This manipulation gives insight into the effect of input of certain variables. This aspect of industrial geology has received increasing attention in recent years from both academicians and practitioners alike. Thus, the book by Paul Martinez and John Harbaugh will add to that body of literature exploring the affects and effects of geological processes. Their goal is to creat realistic simulations in three dimensions.

The authors explore all aspects of sediment transport by waves from constructive to destructive. They give a background into the simulation of waves prior to introducing the reader to computer programs designed to simulate the processes. The operation of the computer program for WAVES and the mathematics behing the algorithm are explained followed by instructions on how to use the program. The program, SEDSIM, is introduced next and instructions on how to use this program are given. They explore then how WAVES and SEDSIM interact and the results of this interaction to form deltas. After describing how to form constructive features, they proceed to explain the destruction of deltas by waves. They give the simulation results of five different situations. All of this is followed by conclusions and projections of use in the future.

Part of the material presented represents dissertation material of Martinez performed under the direction of Harbaugh at Stanford University. The entire project, SEDSIM, is supported by a consortium of petroleum companies and computer vendors. Numerous publications have resulted from this project and some of the results, plus many new ones, have been assembled here in one place for convenience of the reader.

Simulation has been slow to be utilized by the geological profession, but Martinez and Harbaugh, who are at the forefront of the subject, nicely show how it can be used effectively. The main advantage of mathematical modeling as they point out is the incorporation of time and scale into consideration which is not possible with physical models. No doubt additional and expanded applications will result from this work when others see the possibilities. Harbaugh was in the field of simulation early, publishing a landmark book on the subject with coauthor, Graeme Bonham-Carter in 1970. With current interest in basin modeling and analysis, it is possible that simulation of geological processes will come into its own during the decade of the '90s.

Here then is another advance in the subject showing the status-of-the-art as of the early 1990s. The book is sure to be used in the classroom as well as in the 'field'. Let the reader indulge and enjoy.

D. F. Merriam

Preface

The seed for this book was planted in 1985, although its roots extend back much earlier. In 1985 Paul Martinez began graduate study at Stanford, and as part of a class requirement he began a project involving simulation of wave processes. While the computer model developed for the class was simple and two-dimensional, it performed well and it was clear that it could be extended to create more advanced models. The project evolved into a master's thesis completed in 1987 and several publications in turn. Later when Martinez returned to Stanford for a PhD, his dissertation plan was to extend the two-dimensional model to more advanced three-dimensional models. This book, then, is a direct outgrowth of this sequence of events, for it utilizes material from both the master's thesis and the PhD dissertation, although it focuses mostly on the three-dimensional models developed and applied as part of the dissertation.

The more distant roots of this book extend to the period from 1965 to 1969, when initial work on geologic process simulation was begun at Stanford. This early work culminated in a series of publications, most notably in a book published in 1970 by Harbaugh and Bonham-Carter entitled *Computer Simulation in Geology*. Part of the book surveys then-existing and potential applications for computer simulation, and the rest pertains to simulating sedimentary processes as components of integrated systems. The book's general message is that simulating geologic processes is feasible and that there are major advantages in dealing with geological processes as components of integrated dynamic systems.

This point of view remains, and the present book remains in accord with this early philosophy. We emphasized then and we emphasize now, that geological features need to be treated as components of dynamic systems. After all, the earth is a complex dynamic system and its individual components should not be isolated from the whole. While wave processes form only a small part of the spectrum of processes that have created the earth, they are locally important in shaping coastal features and in creating nearshore deposits. Our goal is to show how wave processes can be represented and linked with other processes in integrated systems that permit wide-ranging experiments to be performed.

ACKNOWLEDGMENTS

We are grateful for the sponsorship of the SEDSIM (SEDimentary Basin SIMulation) Project at Stanford University by a consortium of oil companies that includes Agip, Amoco, British Petroleum, Conoco, Elf Aquitane, Marathon, Mobil, Phillips, Shell, Statoil, Japan National Oil Company, and Texaco. The work presented here is a component of the overall SEDSIM project.

Texaco funded initial development of SEDSIM before the consortium was established, and Donald Beaumont, Martin Perlmutter, Richard Byrd, and Michael Zeitlin of Texaco were influential in securing Texaco's support. Bill Goodmen of ARCO Oil & Gas Company helped arrange funding by ARCO for a two-year scholarship for Martinez at Stanford. Silicon Graphics, Gould, Ardent, Stardent, Apple, Interactive Concepts Incorporated, and Dynamic Graphics provided hardware, software, and technical support for computing and graphic display. Robert Sampson of the Kansas Geological Survey wrote and provided SURFACE III, a computer program used for generating and plotting contour maps and other graphic displays. Color plates showing three-dimensional displays employed program SEDVIEW provided by Dr. Christoph Ramshorn (Ramshorn and others, *in press*; and Pflug and others, 1992), and IVM (Interactive Volume Modeling) provided by Dynamic Graphics.

Professors Stephan A. Graham and James C. Ingle of Stanford University, and Dr. H. Edward Clifton of Conoco, consulted on much of the work presented here. Dr. John R. Dingler of the U.S. Geological Survey reviewed early manuscripts dealing with two-dimensional simulation of wave processes, and suggested improvements in computing procedures. Dr. Robert A. Dalrymple of the University of Delaware provided computer programs for modeling wave processes that were modified and incorporated in our programs that simulate wave processes and are collectively known as "WAVE". Professor Rudy Slingerland of Penn State University worked with the SEDSIM group at Stanford during a half-year sabbatical in 1989 and was an important source of ideas. Much of the work involving sediment transport in the surf zone is based on studies by Professor Paul D. Komar, of Oregon State University, and we are thankful to him for arranging a visit at his institution and for serving as a helpful correspondent.

This work has benefitted from colleagues presently or formerly at Stanford. Dr. Daniel M. Tetzlaff's early work while at Stanford provided critical components for this work and pioneered some of the methods used here for graphic display. Dr. Young H.

Lee wrote and maintained the version of SEDSIM's code utilized here, and helped link WAVE with SEDSIM's other programs. Dr. Christoph Ramshorn and Dr. Richard Ottolini wrote programs SEDVIEW and SEDSHO for displaying simulation results as dynamic, three-dimensional, color images and adapted these programs for Ardent and Silicon Graphics workstations. Dr. Ramshorn also provided the user-interface for INTERACTIVE WAVE, helped link WAVE with SEDSIM, reviewed the manuscript, and provided helpful suggestions. Luis Ramos-Martinez provided suggestions for optimizing WAVE's procedures for solving equations involving numerical solutions. Dr. Klaus Bitzer reviewed early manuscripts describing two-dimensional simulation of wave processes and offered ideas and suggestions regarding development of WAVE's three-dimensional procedures. Johannes Wendebourg helped with the development of INTERACTIVE WAVE and, along with Dominik Ulmer, improved input-output procedures for generating contoured graphic displays and provided suggestions for improvements to WAVE and SEDSIM.

PAUL A. MARTINEZ   JOHN W. HARBAUGH

CHAPTER **1**

Introduction

Coastlines are complex and fragile depositional environments. Man's activities can disturb the equilibrium of coastlines by altering local wave climates, changing rates of sediment transport, and changing paths of rivers. In recent decades, engineers have tried to predict nature's reaction to changes in coastal environments by using computer procedures that combine empirical relationships and classical hydrodynamic wave theory. These procedures have proven useful, but are usually very simplified. Historically, they have yielded only static predictions, have not provided three-dimensional records of sediments deposited on coastlines, and have had only limited success as predictive tools over short distances and small spans of time. The objective here is to provide computer procedures that realistically represent nearshore processes and can supplement or replace trial-and-error methodology. These procedures simulate sediment transport by waves and fluvial processes on beaches and deltas at various scales. They should aid coastal engineers, oceanographers, and sedimentary geologists who focus on both modern and ancient nearshore deposits.

How do you simulate nearshore processes using a computer? Can evolving deltaic and coastal environments be simulated realistically by mathematically representing the physical processes that create them? And, once the physics and mathematical formulation are described, what are techniques for transforming them into computer programs? We deal with all of these aspects and take a "how to" approach in guiding the reader through development of computer models for simulating sediment transport in coastal environments. However, this monograph is not intended solely to describe the computer programs that we have devised.

1

Instead, our main purpose is to provide ideas for those who formulate their own mathematical models for simulating nearshore processes.

Simulation of geologic processes is a young science. While scientists and engineers may be adept at formulating computer procedures that provide static solutions, they often lack experience in formulating dynamic procedures that attempt to mimic the behavior of evolving coastal environments. Thus, many mathematical models have been formulated to represent nearshore processes, but few are dynamic, and while static solutions may be useful, they cannot adequately represent transport of sand along a beach or the influence of waves on a growing delta. We have developed simulation models collectively called "WAVE" to represent dynamic geologic processes operating in three dimensions and evolving through time. Results of experiments are displayed with three-dimensional color displays viewed in motion in video-like form for comparison with actual nearshore environments. Plates 1, 2, and 3 and Figures 1-1 to 1-3 provide a preview of experiments of beaches and deltas provided in the following chapters. WAVE stands alone in its class in that it involves multiple grain types, provides a historical record of ages and compositions of deposits, operates at a variety of scales, simulates the passing of a few seconds or a few thousand years, simulates sediment transport by such widely different processes as waves or rivers, and displays results in motion using three-dimensional graphic workstations.

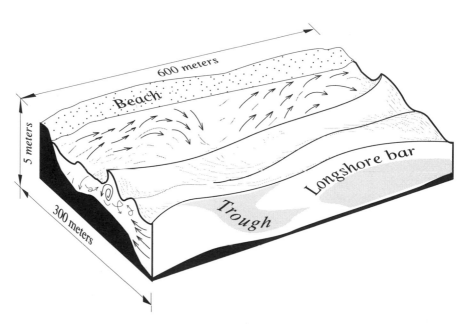

Figure 1-1 Principal components of nearshore environments, including beach, longshore bar, trough, and wave-induced currents. Example represents beach along Oregon coast described by Hunter and others (1979).

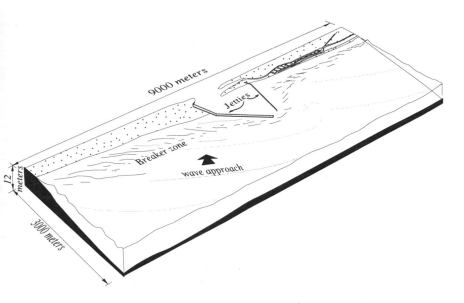

Figure 1-2 Beach with jetties at Anaheim Bay, California described by Caldwell (1956).

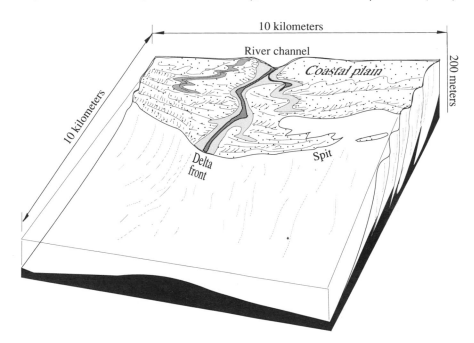

Figure 1-3 Reconstruction of wave-dominated delta of Cretaceous age in south-central Texas based on well data. After Weiss (1980).

3

Graphic display of experimental results is a vital component of computer procedures because information not displayed remains largely uninterpreted and unused. Accordingly we describe procedures and requirements for graphic display that affect storage techniques, formats of data files, and the frequency that results are written to external files. Displays employed by WAVE show successive time steps of experimental results and employ various kinds of two- and three-dimensional displays to monitor results (Plates 4 and 5). Contour maps display nearshore topography, with arrows showing directions of currents and colors indicating sediment composition and transport rates (Plates 8 and 9).

Methodology is also important in devising computer procedures. While many equations have been devised to describe nearshore processes, few published examples document their transformation into usable computer procedures. As experienced computer modelers know, describing complex natural features with mathematical approximations requires creativity and the ability to transform seemingly intractable equations into solvable components (Figure 1-4). This ability is important in modeling nearshore processes that are too complex and poorly understood to be represented by any single set of equations. Accordingly, we describe the judgments and assumptions incorporated into computer programs described here. While there is never a single "correct answer", we have attempted to justify our choices while also showing alternatives. Therefore, our intent is not merely to develop models from existing sets of equations, but also to show how equations themselves are derived and translated into usable computer code.

PREVIOUS WORK

Systematic study of nearshore processes began in the 1940's and 50's when equations based on classical wave theory introduced a century earlier by Airy (1845) and Stokes (1847) were used to predict wave-induced currents and sediment transport. These equations represent many characteristics of waveforms, but equations that relate sediment transport to water waves were more difficult to devise. Workers including Duboys (1879), Bagnold (1940, 1956, 1962, 1963, 1966), Meyer-Peter and Muller (1948), Einstein (1948, 1950, 1972), Watts (1953), Caldwell (1956), Inman and Bagnold (1963), Komar and Inman (1970), Bijker (1976), and Madsen and Grant (1976) studied sediment transport and proposed various transport equations that are now used in simulation models of nearshore and fluvial environments.

Only in the last two decades, however, have workers tried to formulate computer models that dynamically simulate the evolution of nearshore environments through time. Fox (1985) reviewed physical, statistical, probabilistic, and deterministic models used to represent different aspects of nearshore environments, and Lakhan (1989) organized a symposium volume dedicated to a discussion of nearshore processes. Horikawa (1988) and Seymour (1989) reviewed nearshore simulation procedures and documented that while numerous computer models

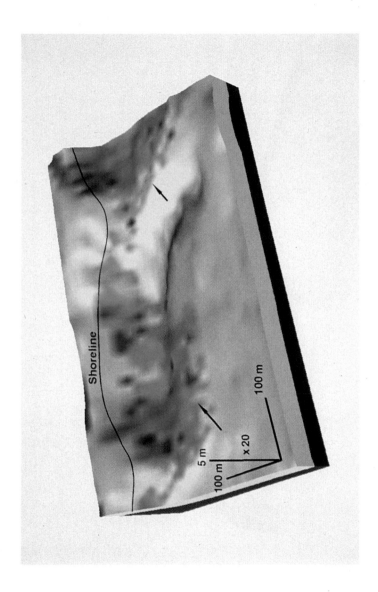

Plate 1 Perspective display of Experiment 9 involving beach with nearshore bar and rip channel similar to setting shown schematically in Figure 1-1. Colors represent a range of grain sizes between medium sand (red) through fine sand (blue). Experiment spanned 24 hours. Fine sediments have been preferentially moved, indicated by green and blue in surf zone. Highest parts of submerged bars shown by arrows. Bold black line shows position of shoreline. Vertical scale is 20 times horizontal. Dimensions of block are 300 m by 600 m by 5 m.

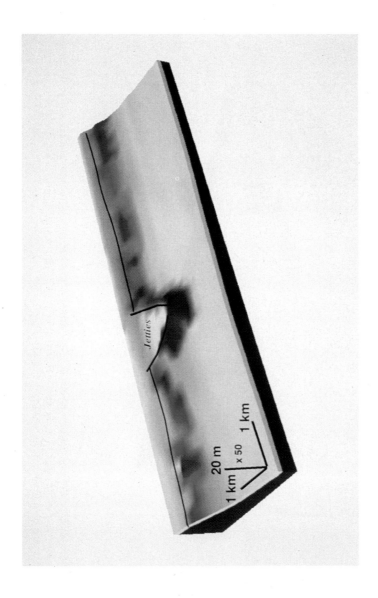

Plate 2 Perspective display from Experiment 8 involving jetty at Anaheim Bay, California observed by Caldwell (1956), and introduced schematically in Figure 1-2. Experiment spans one year. Location of surf zone denoted by different colors, where differences represent differences in sediment composition caused by longshore transport. Finest sediments are shown by blue and coarsest by red. Bold black line shows position of shoreline and jetties. Dimensions of block are 3 km by 9 km by 20m.

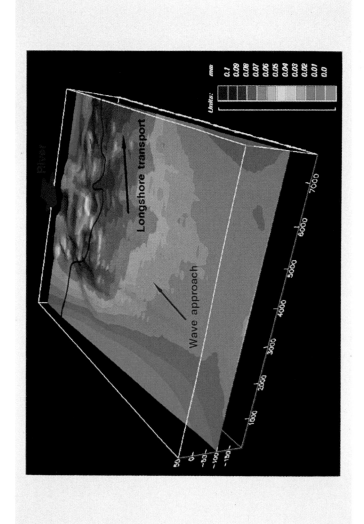

Plate 3 Perspective display of Experiment 11 showing delta affected by waves after 1000 years, and introduced schematically in Figure 1-3. Finest sediments are shown by blue and coarsest by red. Coarsest sand (red and orange) is concentrated near shore and moved towards right by longshore transport (bold black arrow). Red arrow shows path of river. Vertical scale is eight times horizontal. Dimensions of block are 7.5 km by 10 km by 250m.

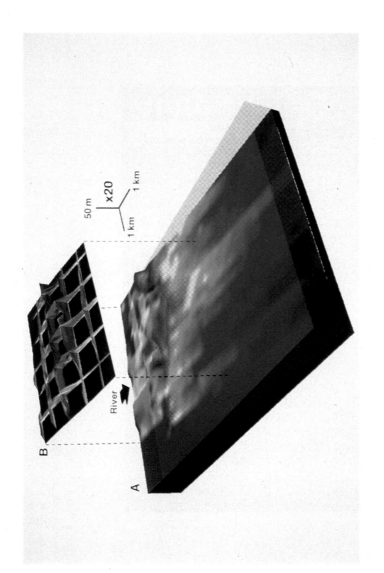

Plate 4 Perspective displays of wave-dominated delta of Experiment 11 after 500 simulated years, showing three-dimensional record of simulated deposits. Red, orange, yellow, green, and blue represent end members consisting of fine sand, very-fine sand, coarse silt, and fine silt, respectively. Vertical scale is 20 times horizontal: (A) Topographic surface where colors represent sediment composition at sediment-water interface. Note concentration of coarsest materials near shore. Black arrow shows path of river. (B) Fence display in which colors denote composition of sediments. Fence sections are 1 km apart.

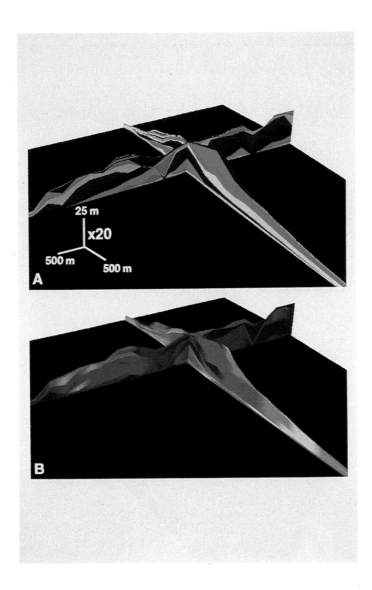

Plate 5 Perspective fence display of wave-dominated delta of Experiment 11 after 500 years, showing selected fence sections of ages and composition of rocks after 500 years. Vertical scale is 20 times horizontal: (A) Sections showing age of beds as bands of color, each band representing deposits formed during a 50-year interval. (B) Composition of beds. Coarsest sand is denoted by red and orange and was concentrated towards shore (right) by longshore transport.

Plate 6 Experiment 1- Beach with constant slope. Perspective display showing composition of sand at surface of beach after twelve iterations, during which time bedload of medium sand moved a distance of twelve grid cells, in direction of red arrow. Conditions of experiment are schematically described in Figure 5-17. Very fine sand (blue) covers west half of beach, whereas medium sand (yellow) covers east half. Location and width of surf zone are denoted by discoloration caused by sorting and mixing of sediment by longshore transport. Waves approached from lower-right corner. Black line denotes position of shoreline. Vertical scale is 10 times horizontal. Dimensions of block are 250 m by 500 m by 6 m.

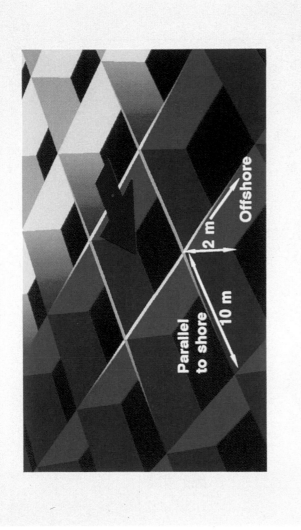

Plate 7 Experiment 1- Beach with constant slope. Enlarged fence display from area near arrow in Plate 6, showing sections parallel and perpendicular to shore after two iterations. Bedload is visible as thin layer of medium-grained sand (yellow) above finer sand (blue). Distance between sections is ten meters, thickness of moving bedload is four centimeters, and thickness of total sand fill was two meters at outset of experiment. Waves approached from right, moving bedload alongshore (red arrow) towards left corner. Vertical scale is 2.5 times horizontal.

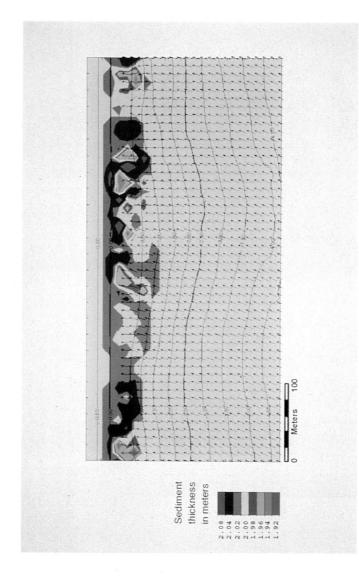

Plate 8 Experiment 2- Beach with irregular slopes. Map shows depths with contours in meters, wave orthogonals with arrows, and sand thickness with colors. Waves approached from lower-left corner. Red and orange areas are up to a few centimeters thicker than the two-meter thickness of sand (yellow) present at outset of experiment. Wave orthogonals and sand thickness are shown on cell-by-cell basis. Grid representing area has 25 rows and 50 columns containing 1250 grid cells, each 10 meters square.

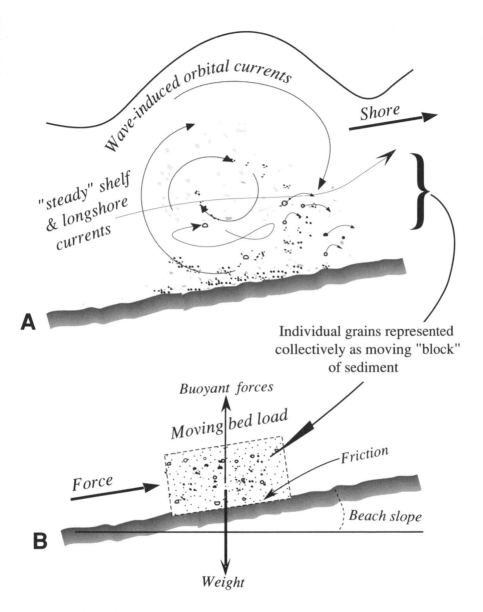

Figure 1-4 Sediment transport can be represented by sediment grains moving as cohesive volume of sediment influenced by wave-induced forces, and by friction, gravity, and buoyancy: (A) Oscillatory and steady wave-induced currents transport sediment both as bedload and suspended grains. (B) Computer procedures represent moving grains as cohesive volume.

5

exist for simulating currents in nearshore systems, only a few simulate sediment transport caused by nearshore currents.

Aspects of work presented here began in the 1960's, when simple two-dimensional simulation models of shallow-marine sedimentation were developed at Stanford by Bonham-Carter and Sutherland (1968) and Harbaugh and Bonham-Carter (1970). Tetzlaff (1987) extended their work by developing a three-dimensional simulation model for simulating clastic sedimentation associated with flow in open-channels. His computer model consists of many subprograms collectively called SEDSIM (after the project's name SEDimentary Basin SIMulation), which have been altered and expanded to form SEDSIM's system of programs as they presently exist. Tetzlaff and Harbaugh (1989) used SEDSIM to simulate sedimentary deposits formed by deltas, meandering rivers, braided streams, alluvial fans, submarine fans, and submarine channels. Other applications of SEDSIM's computer programs are described by Martinez (1987), Martinez and Harbaugh (1989), Martinez (1992a, 1992b), Scott (1987), Koltermann and Gorelick (1990a, 1990b), Lee (1991), Lee and Harbaugh (1992), Wendebourg (1991), and Wendebourg and Ulmer (1992).

WAVE links with SEDSIM's other programs to form a unified model that simulates sediment transport caused by waves and rivers (Figures 1-5 and 1-6). Together they probably represent the most complete and rigorous procedures for simulating sediment transport by waves and rivers. WAVE does not simulate biological or chemical processes, although these processes may be locally important and could be incorporated in subsequent extensions of WAVE.

TYPES OF SIMULATION MODELS

Two general types of models that simulate physical geological processes are in general use, namely (1) physical models such as wave tanks, that are small-scale replicas of larger systems and (2) mathematical models that represent processes using equations and logic operations, generally in the form of computer programs that execute operations repeatedly. Physical models are sometimes desirable because they directly represent actual systems or processes, but their application is usually restricted by their small size. For example, it is difficult or impossible to recreate the influence of large breaking waves, rip currents and longshore currents in a laboratory model. By contrast, mathematical models generally do not have such scaling problems, but they invariably must be simplified so that processes can be represented by equations and logic operations.

Mathematical models also allow more versatility in controlling of experiments. Wide-ranging experiments can be attempted, allowing different hypotheses to be tested and providing alternative interpretations. For example, simulation of the delta shown in Plates 3, 4, and 5 provides much more information than the purely conceptual cartoon in Figure 1-3. Furthermore, the computer model could be run iteratively until simulated deposits shown by cross sections in Plate 5

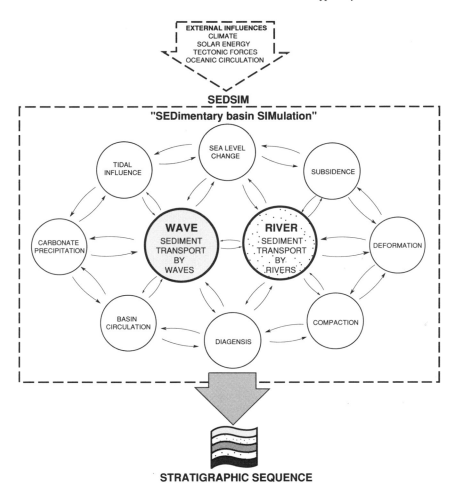

Figure 1-5 Major component modules for simulating sedimentary basins, that include linkage of WAVE and rest of SEDSIM for simulating sediment transport by waves and rivers. Arrows represent interdependence between modules.

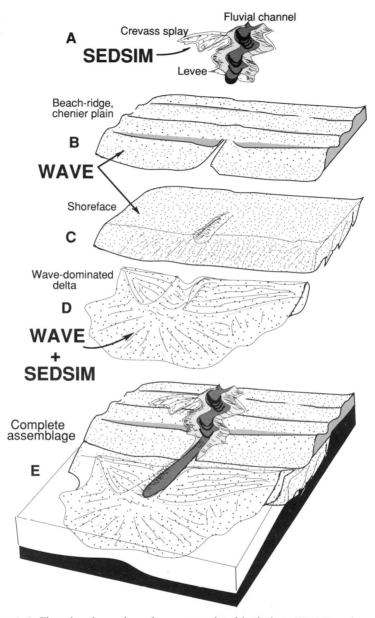

Figure 1-6 Fluvial and nearshore features simulated by linking WAVE and rest of SEDSIM: (A) SEDSIM simulates fluvial system including levee and crevass-splay deposits. (B) WAVE reworks fluvial sediments into beach ridge and chenier plain deposits. (C) WAVE forms shoreface deposits by simulating littoral transport of sand. (D) WAVE and SEDSIM together simulate deltaic sands reworked by waves. (E) Assemblage forms wave-dominated delta.

compare favorably with subsurface stratigraphic features interpreted from well-logs and seismic sections. Specific parameters also could be changed while holding others constant, providing insight about the influence of different variables. Or experiments could contrast a wave-dominated delta with a fluvial-dominated delta, allowing comparison of complex delta systems that previously could only be observed in the field by several generations of workers.

Tetzlaff and Harbaugh (1989) provide a simplified classification of mathematical simulation models. Models are typically either deterministic or probabilistic, static or dynamic, and zero-, one-, two-, or three-dimensional. WAVE and SEDSIM's other programs are deterministic, dynamic, and three-dimensional. There are no random components in deterministic models, and unlike random models, deterministic computer models should produce identical results when provided with identical input data and operated with the same computer and operating software.

If time and change are represented in a mathematical model, it is dynamic. Dynamic models incorporate feedback so that processes operating through time respond to successive changes by simultaneously interacting with each other, allowing an observer to control critical variables and obtain insight about the origin of sedimentary features produced by interactions between flow, sediment transport, and submerged topography. For example, we can combine subprograms that incorporate feedback procedures into a process-response model that simulates the effects of waves impinging on a coastline (Figure 1-7).

Computer models necessarily are simplified versions of actual systems. As Tetzlaff and Harbaugh (1989) describe, "loss of reality" is inevitable and procedures must be developed to ensure that simplifications in the model do not result in unduly unrealistic performance. For example, representing a span of time as a series of discrete steps, or an area with arrays consisting of discrete cells, involves approximations of features that are continuous in both time and space. These approximations are sensitive to the duration of time steps or sizes of cells. Although representation can be progressively improved with shorter time steps and smaller cells, there are tradeoffs because as time steps and cells become smaller, more computational effort is required to simulate a given amount of time and more cells are required to represent a given area.

WAVE'S GOALS

The plan for a computer simulation model should define its purpose and establish criteria for measuring its success. It should consider who will use the model, the kinds of responses desired, the processes of interest, the precision required, and scales at which processes operate. WAVE, when linked with SEDSIM's other programs, is designed to:

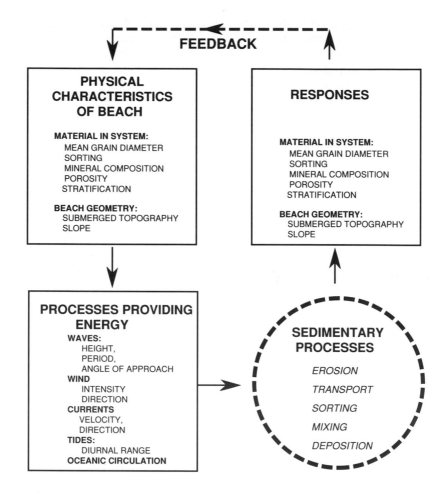

Figure 1-7 Simplified flow diagram of conceptual process-response model of beach. Modified from Krumbein (1964) and Harbaugh and Bonham-Carter (1970).

(1) Simulate nearshore sediment transport by waves in three dimensions, using input data based on and readily available from field observations.

(2) Link with SEDSIM (Figures 1-5 and 1-6) for simulating effects of wave-induced currents on deltaic systems.

(3) Operate over periods ranging from a few seconds to thousands of years.

(4) Represent geographic areas ranging from beaches extending over a few hundred meters, to deltas extending over tens of kilometers.

(5)   Provide quantitative estimates of wave parameters that include heights of breaking waves, refraction angles, and velocities of longshore currents.

(6)   Record and display ages of deposits and provide a detailed record of erosional and depositional events.

(7)   Represent clastic sediment composed of different grain sizes and densities, such as sand, silt, and clay.

(8)   Record and display the composition of deposits of varying grain sizes.

(9)   Employ three-dimensional, interactive, color, video-like graphic displays.

(10) Provide insight in the study of modern and ancient nearshore environments.

ASSUMPTIONS IN WAVE

Before a computer program can represent nearshore processes, we must understand how those processes operate and represent them at some suitable level of detail. We assume that coastlines can be treated as a hierarchical system of depositional environments (Figure 1-8) in which physical processes can be represented at progressively smaller scales, such as those that involve breaking waves or movement of fluid in an open-channel. At a more detailed level, interactions between fluid and sediment can be represented. This divide-and-conquer approach simplifies development of computer programs because their organization and linkage can parallel the hierarchical organization of natural systems where smaller, interdependent computer programs representing various smaller processes (Figures 1-5 and 1-6) are linked to represent larger natural systems.

Figure 1-9 shows major factors that affect a hypothetical coastline where a river supplying sediment has created a delta. Solar energy, the earth's rotation, and heat from the earth's core are the major driving forces for the most influential factors which include (Coleman and Wright, 1975) (a) climate, (b) topographic relief in the drainage basin, (c) the river's fluid discharge rate, (d) the river's sediment discharge rate, (e) whether the river's flow is homopyncnal, hyperpyncnal, or hypopyncnal relative to the water in the receiving basin (Bates, 1953), (f) the nearshore wave climate, (g) wind, (h) nearshore currents, (i) tidal energy, (j) slope of the receiving basin, (k) submerged topography of the receiving basin, and (l) tectonic deformation and isostatic subsidence of the receiving basin. Wright and Coleman (1973) conclude that no single factor plays a greater role in coastline development than does the wave regime.

Not all of these factors, however, are represented in WAVE and SEDSIM, which can simulate only certain aspects of wave, and fluvial processes. WAVE and SEDSIM assume that climate, morphology, geology, and soil in an alluvial system

11

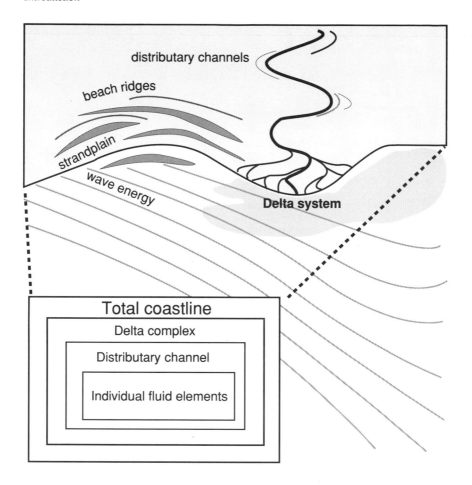

Figure 1-8 Diagram showing nesting of systems ranging downward in scale from total coastline, delta complex, distributary channels, to individual fluid elements. Large depositional systems can be represented by simulating components at different scales.

are represented only indirectly by sediment grain sizes, sediment discharge rates, and fluid discharge rates provided as input. Similarly, wind direction, magnitude, and fetch are represented indirectly by providing appropriate wave heights, lengths, periods, and angles of approach.

Processes not represented in WAVE and SEDSIM include tidal currents, oceanic circulation, compaction, diagenesis, subsidence, carbonate precipitation, and eolian transport, although they could be added later (Figure 1-5). Furthermore, WAVE does not represent shelf currents, although they may be important for transporting sediment suspended by waves. Seasonal or catastrophic events including storms, hurricanes, or flood are not directly represented, although they could be indirectly represented by periodically changing input data to represent unusually large waves or river discharges.

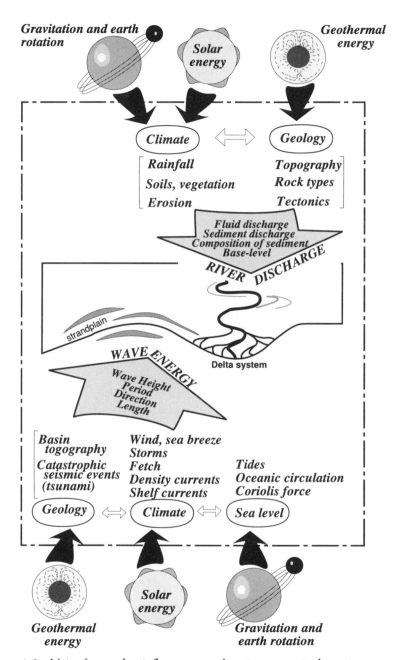

Figure 1-9 Major factors that influence coastal environments in dynamic system are driven by solar energy, earth's rotation, and heat from earth's core. Dashed border defines arbitrary system boundary in simulation models, where solar, geothermal and rotational influences are external but affect processes inside boundary.

In linking WAVE with the rest of SEDSIM to represent wave and fluvial processes, we have made the following general assumptions:

(1) A typical coastal system such as that shown in Figure 1-9, consists of deposits formed dominantly by waves and rivers.

(2) Coastal systems can be represented by simulating processes that operate at smaller scales (Figure 1-8).

(3) Littoral transport by waves can be represented by equations that relate empirical equations for sediment transport to wave hydrodynamics.

(4) Fluvial processes in deltas can be represented by equations for open-channel flow.

(5) Catastrophic storms or hurricanes are not included, but effects of large waves can be represented by providing WAVE with large wave heights as input.

(6) Chemical and biological processes are not represented in WAVE, although they may be locally important and could be included later.

(7) WAVE represents unconsolidated sediments consisting of sands, silts, and clays whose densities may vary.

(8) Sediments have 40 percent porosity when deposited, but post-depositional alterations including cementation, compaction, and deformation are not represented.

(9) Both WAVE and SEDSIM are "quasi three-dimensional" in their representation of flow because they employ "depth-averaged" currents to represent currents that are averaged so that at any geographic location, the velocity is uniform vertically over the depth of flow. Furthermore, WAVE employs "time-averaging", where currents are averaged during an interval of time equal to the wave period.

(10) The use of time-averaged and depth-averaged currents limits WAVE's ability to represent oscillatory, orbital motions that cause onshore and offshore transport, so that cross-shore sand transport that is dominant seaward of the breaker zone is not fully represented.

(11) WAVE is best suited for simulating sediment transport in the surf zone, located between the breaker zone and shore, where steady longshore currents are the primary mechanism for transporting sand.

14

(12) Sand transport by shelf and tidal currents and subaerial transport by wind are not represented, but could be represented later.

(13) Coriolis effects and effects of oceanic circulation are assumed to be negligible at the scale at which WAVE operates and are not represented.

(14) The performance of WAVE and SEDSIM can be adequately evaluated by comparing results of experiments with published field data.

DEVELOPMENT OF WAVE

The sequence of steps involved in developing WAVE is illustrated in Figure 1-10. Each major step is described in a chapter whose number accords with the numbered steps summarized in Table 1-1. Chapter 1 provides an overview of

Table 1-1 Seven major steps used in developing WAVE, each of which accords to Chapters 1 through 7, schematically shown in Figure 1-10.

Step 1 (Chapter 1)	Define goals, assumptions, and sequence of steps involved in developing computer procedures for simulating nearshore processes.
Step 2 (Chapter 2)	Describe physical features of nearshore environments and define processes incorporated in computer programs.
Step 3 (Chapter 3)	Select equations for water-wave mechanics and represent them with finite-difference approximations.
Step 4 (Chapter 4)	Select equations for sediment transport and represent them with analytical approximations.
Step 5 (Chapter 5)	Devise procedures that link computer programs for simulating wave hydrodynamics and sediment transport, and devise accounting and graphic display schemes that represent ages and compositions of simulated deposits.
Step 6 (Chapter 6)	Conduct simulation experiments for comparison with field data from modern beaches to test applicability of procedures to various environmental situations and geographic dimensions.
Step 7 (Chapter 7)	Link WAVE with the rest of SEDSIM to conduct simulation experiments of large-scale ancient features including ancient deltas and coastlines.

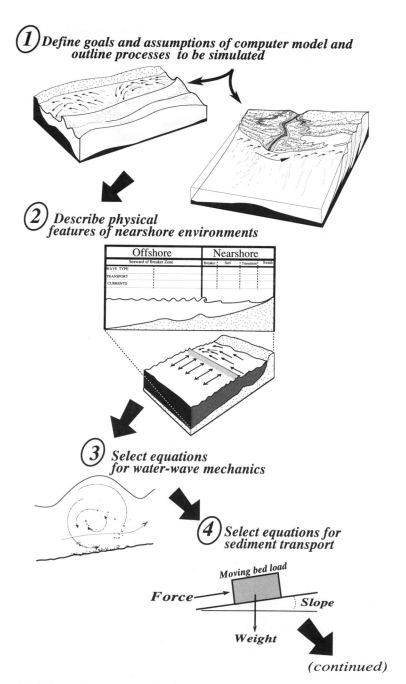

① *Define goals and assumptions of computer model and outline processes to be simulated*

② *Describe physical features of nearshore environments*

③ *Select equations for water-wave mechanics*

④ *Select equations for sediment transport*

(continued)

Figure 1-10 Seven major steps in developing WAVE accord with Chapters 1 through 7.

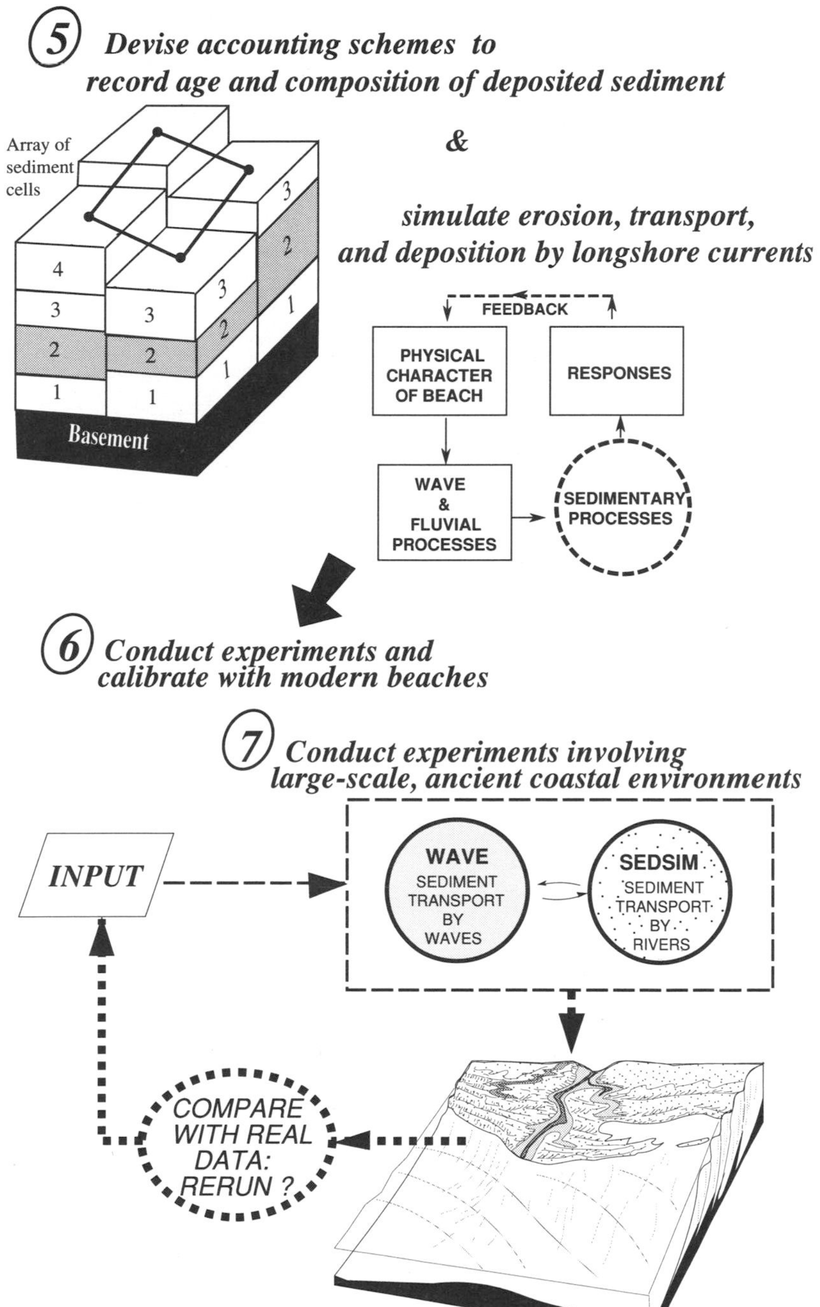

⑤ *Devise accounting schemes to record age and composition of deposited sediment*

Array of sediment cells

&

simulate erosion, transport, and deposition by longshore currents

4
3
3
2
1

3
2
1

3
2
1

Basement

FEEDBACK

PHYSICAL CHARACTER OF BEACH

RESPONSES

WAVE & FLUVIAL PROCESSES

SEDIMENTARY PROCESSES

⑥ *Conduct experiments and calibrate with modern beaches*

⑦ *Conduct experiments involving large-scale, ancient coastal environments*

INPUT

WAVE
SEDIMENT TRANSPORT BY WAVES

SEDSIM
SEDIMENT TRANSPORT BY RIVERS

COMPARE WITH REAL DATA: RERUN ?

Figure 1-10 (continued)

17

simulation models and our goals in developing WAVE while Chapter 2 further defines the scope of this monograph by describing physical features, processes, and environments that are represented by WAVE.

In developing WAVE, we focused on three major tasks separated accordingly in Chapters 3, 4, and 5. In Chapter 3 we describe the "behavior of the water" by describing the physics of shoaling water waves, their role in creating nearshore currents, and the mathematical approximations used to describe them. In Chapter 4 we describe the "behavior of the transported sediment" by describing the physics of sediment transport caused by waves, and the methods used to calculate rates of sediment transport. In Chapter 5 we present accounting methods used in the computer programs to represent time, space, and an evolving stratigraphic record, and describe how ideas and equations are transformed into subprograms and linked together to form a unified computer simulation model.

Chapter 6 presents experiments where the simulation models were tested with data from modern beaches that vary widely in size, with lengths from hundreds to thousands of meters. These experiments may interest engineers who require quantitative estimates involving coastal or beach systems that change during periods spanning days, weeks, months, or a few years. By contrast, Chapter 7 presents experiments involving larger deltaic systems whose simulations span tens or hundreds of kilometers, and hundreds to thousands of years. These experiments should interest geologists who focus on the three-dimensional arrangement of coastal deposits forming ancient stratigraphic sequences. Finally, Appendix A provides data used for calibration of and comparison with experiments. Appendices B, C, and D provide descriptions of input data provided to WAVE and SEDSIM, and Appendix E provides definitions of symbols used in the text.

CHAPTER 2

Physical Features and Processes of Nearshore Environments

This chapter describes features of nearshore environments that are suitable for simulation with WAVE and the rest of SEDSIM, where WAVE simulates sediment transport by waves and SEDSIM simulates sediment transport by rivers. Coastlines generally include various sub-environments (Figure 2-1) and may be much more complex than the simple coastline depicted in Figure 1-8, and while WAVE and SEDSIM cannot simulate all processes involved in coastal environments, they do simulate many aspects of processes that create beaches, spits, barrier bars, and deltas. These various environments are described here to provide comparisons with experimental results presented in subsequent chapters. A collective overview of features of nearshore environments is provided by Krumbein (1944), Bascom (1954), Ingle (1966), Komar (1976), Greenwood and Davis (1984), Sleath (1984), Horikawa (1988), the *Shore Protection Manual* (U.S. Army CERC, 1977), and by other authors whose works are cited later.

Nearshore environments can be divided into offshore and nearshore areas and described with terms shown in Figures 2-2 and 2-3. The offshore area lies seaward of the breaker zone, whereas the nearshore area includes the breaker, surf, and swash zones. Offshore and nearshore areas can be further subdivided according to characteristics of incoming waves, mechanisms of sand transport, and patterns of wave-induced currents. Three-dimensional representation of nearshore processes is required to fully represent the directions of sand transport because wave-induced currents vary between the nearshore and offshore areas (Figure 2-2). Onshore-offshore transport results from oscillatory motion of waves, is dominant in the offshore area, and involves motions generally parallel to the direction of wave propagation. By contrast, longshore transport dominates shoreward of the breaker

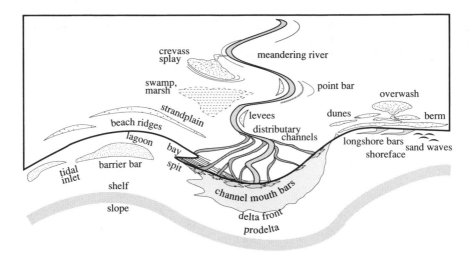

Figure 2-1 Typical onshore, nearshore, and offshore features of coastal systems.

zone and generally involves movement parallel to shore, although rip currents within the breaker zone may move sediment offshore. Thus, currents produced by waves need to be represented in three dimensions to represent patterns of sediment transport caused by combinations of onshore-offshore, longshore and rip currents. Two-dimensional models are incapable of representing all these motions.

Positions of nearshore and offshore areas (Figure 2-4) fluctuate seasonally (Figure 2-5) and even hourly (Figure 2-6) in response to changes in wave climate, tidal phase, and beach slope. An example in Figure 2-5 involves seasonal changes in wave climate and corresponding changes in beach topography. In response to a change in wave climate, the topography of a beach generally moves toward equilibrium, with the shoreline's position and topography adjusting to the new wave climate. To be effective, computer models must represent beaches as they evolve toward equilibrium conditions following a change in wave climate. While two-dimensional simulation models have been widely used to predict changes in beach profiles, including those by Bowen (1969, 1980), Felder and Fisher (1980), Leontev (1985), Bailard (1981, 1984), Seymour and Castel (1989), Martinez (1987a,b), and Swain (1989), WAVE has major advantages because it yields three-dimensional predictions as beaches evolve toward equilibrium.

Offshore areas

Far offshore, waves are generally symmetric, but as they propagate toward shore they become asymmetric with steeper crests and wider troughs (Figure 2-7), producing oscillatory or orbital currents that cause both onshore and offshore transport (Figure 1-4A). Although shelf currents transport sediment suspended by waves in offshore areas, waves in offshore areas are largely neglected because WAVE does not incorporate shelf currents. Instead, WAVE focuses on nearshore

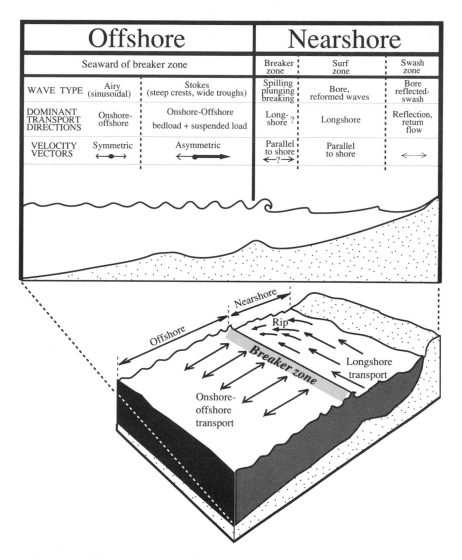

	Offshore		Nearshore		
	Seaward of breaker zone		Breaker zone	Surf zone	Swash zone
WAVE TYPE	Airy (sinusoidal)	Stokes (steep crests, wide troughs)	Spilling plunging breaking	Bore, reformed waves	Bore reflected-swash
DOMINANT TRANSPORT DIRECTIONS	Onshore-offshore	Onshore-Offshore bedload + suspended load	Long-shore ?	Longshore	Reflection, return flow
VELOCITY VECTORS	Symmetric ←—●—→	Asymmetric ←——●——→	Parallel to shore ←—?—→	Parallel to shore	←——→

Figure 2-2 Classification of nearshore versus offshore areas with respect to location of breaker zone. Subdivisions list wave types, sediment transport mechanisms, and wave-induced velocities. Compiled from Ingle (1966) and Clifton and others (1971).

21

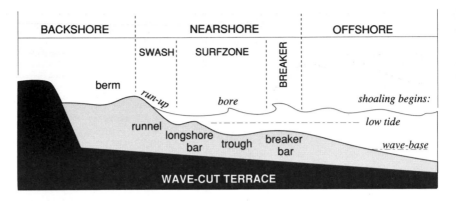

Figure 2-3 Cross section illustrating terminology associated with nearshore environments. Modified from Ingle (1966).

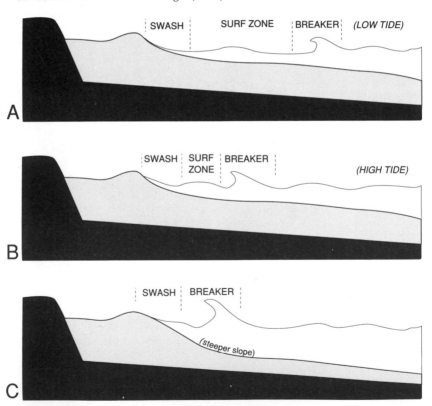

Figure 2-4 Schematic cross sections through beaches showing how width of surf zone is influenced by beach slope, tidal phase, and wave climate: (A) Gently sloping beach at low tide has wide surf zone, whereas same beach (B) at high tide has narrow surf zone. (C) More steeply sloping beach may lack surf zone. Modified from Ingle (1966).

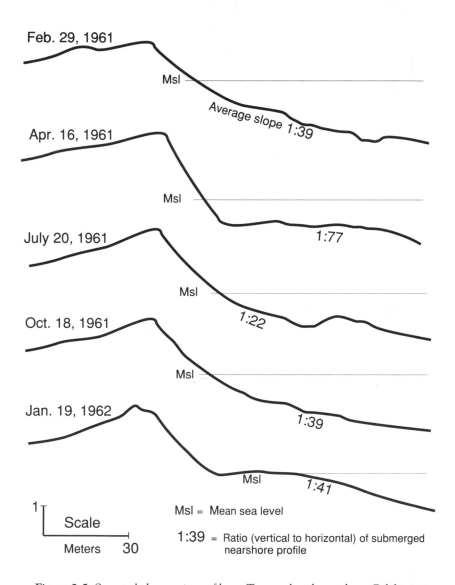

Feb. 29, 1961

Msl

Average slope 1:39

Apr. 16, 1961

Msl

1:77

July 20, 1961

Msl

1:22

Oct. 18, 1961

Msl

1:39

Jan. 19, 1962

Msl 1:41

1
Scale
Meters   30

Msl = Mean sea level

1:39 = Ratio (vertical to horizontal) of submerged nearshore profile

Figure 2-5 Seasonal changes in profiles at Trancas beach, southern California in 1961 and 1962. Ratios pertain to average slope of profile below mean sea level. Vertical scale is ten times horizontal. Modified from Ingle (1966).

23

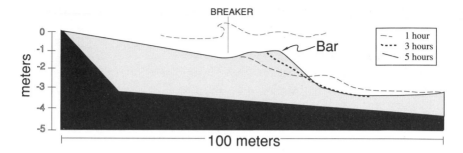

Figure 2-6 Changes in beach profile in two-dimensional simulation experiment involving onshore-offshore sand transport seaward of breaker zone. Profiles represent first, third, and fifth hours after beginning of experiment. Bar formed as sediment progressively moved toward shore (Martinez, 1987a).

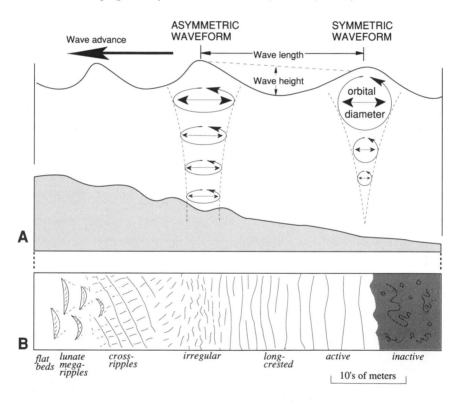

Figure 2-7 Cross section and map showing how symmetric and asymmetric waveforms affect bedforms: (A) Circular motions in deep water correspond to symmetric waveforms. As waves move toward shore they become asymmetric, with elliptical motions that touch bottom and transport sediment both onshore and offshore. (B) Map of bottom corresponding to cross section showing bedforms associated with shoaling waves. Modified from Clifton (1976) and Clifton and Dingler (1984).

transport, although procedures could be incorporated for representing transport by shelf currents.

Nearshore area

As waves move towards shore, water moving along the seabed encounters more friction than along the air-water surface, allowing water high in the water column to move faster. Eventually waves break as water at the air-water surface outruns water moving at the seabed. As they break waves cause turbulence that suspends, sorts, and transports sediments. Much of the energy released by breaking waves drives longshore and rip currents that transport sediment alongshore as "littoral drift" within the surf zone (Figure 2-8). Transport rates and littoral drift are generally greatest within the surf zone, causing sand to move parallel to the shoreline similar to a "river of sand" (Einstein, 1948).

After waves break, some energy continues to propagate toward shore in the form of bores that move toward the swash zone (Figures 2-2 and 2-3). The swash zone is transitional between the sea and shore where incoming bores or reformed waves are reflected back toward the sea. Waves entering the swash zone produce a "surf beat" where pulses of water from broken waves runup on shore and then flow backwards. The swash zone is only intermittently covered by water and generally has a smooth or cusp-like topography. WAVE represents the swash zone and the balance between runup and return flow by employing time-averaged and depth-averaged currents, which are averaged during the passing of one wave period, and are assumed to be uniform through the water column at any specific geographic location.

Submerged topography affects wave-induced currents and sediment transport. Longshore bars, rip channels, and other features affect propagating waves, cause complicated refraction patterns, and produce currents whose directions are influenced by topography (Figure 2-9A). Simultaneously, wave-induced currents shape submerged topography by moving, sorting, and redepositing sediment (Figure 2-9B). The coarsest grains commonly occur in longshore bars formed where wave energy for sorting and reworking is greatest (Figure 2-10). WAVE's currents are likewise sensitive to topography and are recalculated as topography changes, thereby creating a feed-back system characteristic of process-response models (Figure 1-7). WAVE also sorts, erodes, and deposits grains according to their size and density and can represent mixtures composed of up to four grain sizes.

Ripples, sand waves, and other sedimentary structures form both offshore and nearshore (Figure 2-7) in response to waves (Clifton and others, 1971; Clifton, 1976; Dingler and Inman, 1976; Hunter and others, 1979). These smaller bedforms, however, are not represented by WAVE even though WAVE can be applied at geographic scales where they are important. Their simulation would require representation of eddies, vorticity, and other local variations within the fluid column that are not incorporated in WAVE because currents are represented in depth-averaged form in which the velocity is uniform through the total depth at a given geographic location. Furthermore, geographic cells in WAVE are generally

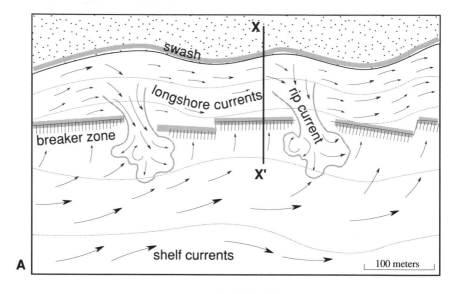

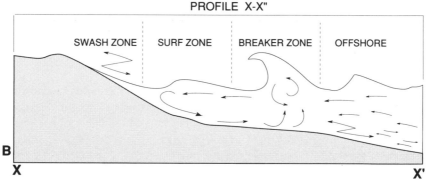

Figure 2-8 Map and cross section of hypothetical beach showing currents induced by shoaling waves: (A) Breaking waves drive nearshore circulation consisting of longshore currents and seaward-flowing rip currents that create circulation cells. Shelf currents seaward of breaker zone are related to basin or oceanic circulation. (B) Cross section X-X' shows wave-induced currents which include onshore and offshore currents beyond breaker zone, turbulence in breaker zone, surface and seaward-flowing bottom currents in surf zone, and run-up in swash zone. Modified from Shepard and others (1941), Shepard and Inman (1950), Ingle (1966), and Horikawa and Sasaki (1972).

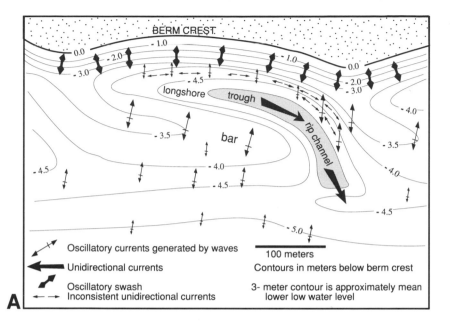

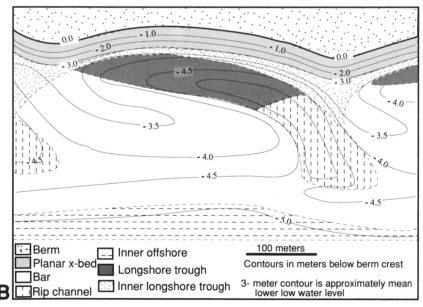

Figure 2-9 Maps of longshore-bar and rip-channel system on southern Oregon coast: (A) Submerged topography and wave-induced currents associated with longshore bar and trough. (B) Classification of features caused by wave-induced currents. Modified from Hunter, Clifton and Phillips (1979).

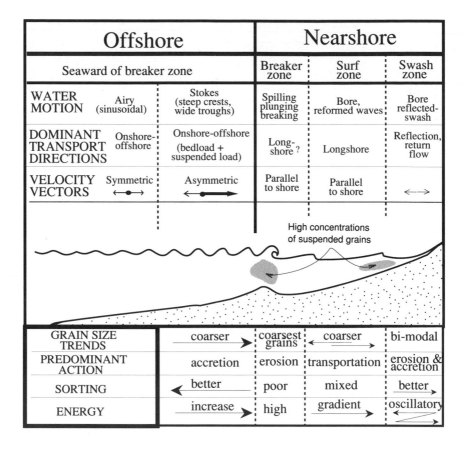

Offshore		Nearshore		
Seaward of breaker zone		Breaker zone	Surf zone	Swash zone
WATER MOTION Airy (sinusoidal)	Stokes (steep crests, wide troughs)	Spilling plunging breaking	Bore, reformed waves	Bore reflected-swash
DOMINANT TRANSPORT DIRECTIONS Onshore-offshore	Onshore-offshore (bedload + suspended load)	Long-shore ?	Longshore	Reflection, return flow
VELOCITY VECTORS Symmetric ←—●—→	Asymmetric ←—●——→	Parallel to shore	Parallel to shore	←—→

High concentrations of suspended grains

GRAIN SIZE TRENDS	coarser →	coarsest grains	← coarser →	bi-modal
PREDOMINANT ACTION	accretion	erosion	transportation	erosion & accretion
SORTING	← better	poor	mixed	better →
ENERGY	increase →	high	gradient →	oscillatory ←→

Figure 2-10 Classification of waves and sediment transport in nearshore and offshore areas. Water motions, directions of currents, and transport directions vary between breaker, surf, and swash zones. Grain sizes and sorting reflect action by waves. Modified from Ingle (1966) and Clifton and others (1971).

much larger than the geographic dimensions of these bedforms, which are therefore beyond the resolution of the cells.

Spits and Barrier Bars

Processes that form beaches also form spits and barrier bars, which can be thought of as large-scale bedforms created by longshore transport of sand (Larue and Martinez, 1989) and can be simulated with the same procedures used for beaches. However, these larger-scale features may be influenced by other factors, including changes in sea level and local subsidence, so their simulation may require additional modules shown in Figure 1-5. For example, experiments in Chapter 7 that involve barriers, spits and deltas typically involve changes in sea level over thousands of years, and would be incomplete without incorporating the effects of sea level fluctuations.

Spits are transient features that may form and then be destroyed by the combined effects of waves, sea level changes, and local subsidence. Spits commonly occur at margins of deltas where rivers provide excess sand to the shoreline. Their internal bedforms (Figure 2-11) reflect lateral migration of sand in the direction of wave movement. WAVE can simulate migrating spits and is potentially useful for engineering studies of coastal erosion involving spits. WAVE also simulates sorting and thus represents some of the processes in the geologic past that created deposits that now serve as hydrocarbon reservoirs. Similarly, barrier bars are also important for their engineering properties and their role as hydrocarbon reservoirs, and are likewise affected by fluvial processes, littoral sand transport, sea level changes, and subsidence, all of which can be represented by linking WAVE with SEDSIM.

EFFECTS OF WAVES ON DELTAS

Deltas form where rivers flow into open bodies of water, creating deposits that build, or prograde beyond the adjacent shoreline. Like beach, spit and barrier-bar deposits, they are important for their engineering properties as well as their role as hydrocarbon reservoirs. Three principal factors that influence deltas (Figure 1-9) include fluvial sediment discharge rate, wave energy, and tidal energy in the receiving basin (Fisher, 1969; Wright and Coleman, 1973). These three factors permit delta types to be classified in a three-end-member spectrum (Figure 2-12) according to relative contributions of fluvial, wave, and tidal processes, but because WAVE and SEDSIM do not include tides, we focus here only on that part of the spectrum involving wave versus fluvial activity.

Deltas strongly influenced by waves are known as "wave-dominated" deltas, and deltas controlled largely by fluvial or tidal processes are "fluvial-dominated" or "tide-dominated" deltas. This classification involves the relative influence of wave, fluvial, and tidal processes that produce "elongate," "lobate," "cuspate" or "estuarine" deltas (Figure 2-12), each of which has distinctive features. Figures 2-13 and 2-14 contrast elongate and cuspate deltas that form end members of the continuum of deltas influenced by wave and fluvial processes.

Fluvial-dominated deltas

Fluvial-dominated deltas form elongate or lobate protrusions from the shoreline and may contain crevass splays, point bars, channels, levees, peat beds, distributary channels, and distributary channel-mouth bar deposits (Figure 2-13). Rivers that form fluvial-dominated deltas generally contribute a relatively high proportion of suspended-load sediment, producing thick prodelta mud deposits. High sedimentation rates and rapid building of delta plains and prodelta deposits produce broad deltaic plains with slopes of less than one degree.

Elongate fluvial-dominated deltas such as the Mississippi delta are characterized by irregular shorelines whose finger-like sand deposits are formed by distributary channels that can be shown by contour maps representing net thicknesses of

29

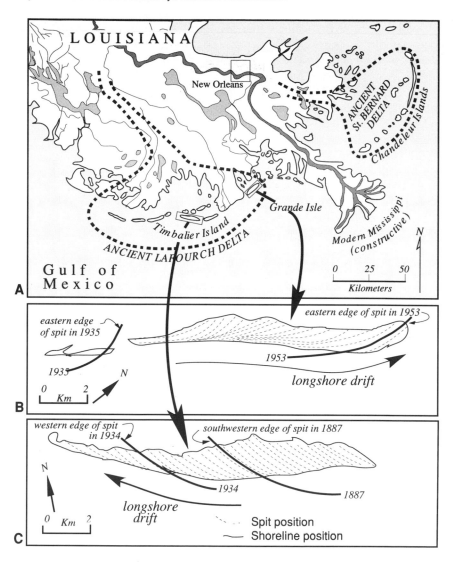

Figure 2-11 Spits associated with Mississippi delta (modified from Penland and others, 1981).(A) Regional map of modern distributaries of Mississippi delta and prominent spits associated with ancient deltas. (B) Enlargement of spit that forms Grande Isle. Bold lines show southeastern edge of spit in 1935 and in 1953, documenting migration. (C) Enlargement of spit that forms Timbalier Island, with bold lines representing southwestern edge of spit in 1887 and 1934.

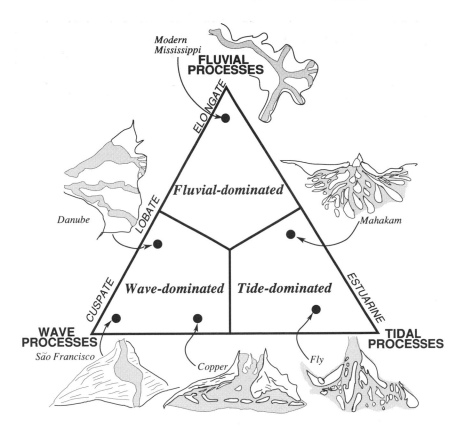

Figure 2-12 Triangular diagram classifying deltas according to domination by fluvial, wave, or tidal processes. Example deltas are located inside triangle with solid circles. Modified from Galloway (1975) and Weise (1980).

sand (Figure 2-13B). Petrophysical logs of boreholes in ancient elongate deltas show upward-coarsening interbedded sand and shale sequences capped by distributary channel and delta-plain deposits (Figure 2-13C). Rapid progradation and channel abandonment produce protruding, finger-like sand bodies that are quickly buried and escape reworking by waves. Lobate deltas such as the Danube delta of Romania (Figure 2-12) reflect increasing marine influence. Lobate deltas have numerous distributary channels and associated fluvial deposits that are similar to elongate deltas, but sands in distributary channels are often reworked into delta-front sheet sands that are thicker and better sorted than those of elongate deltas.

Wave-dominated deltas

Wave-dominated deltas are produced by reworking by waves (Figure 2-14) and are characterized by areally extensive, well-sorted sheet sands aligned parallel to shore. Barrier ridges, cheniers, and beach dunes that are also parallel to shore are

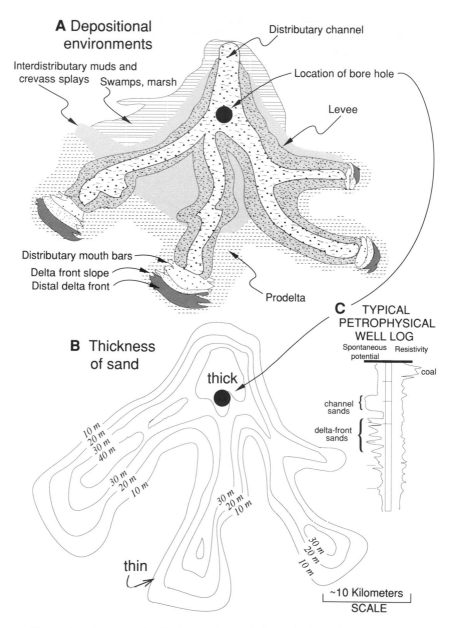

A Depositional environments

Distributary channel

Interdistributary muds and crevass splays

Swamps, marsh

Location of bore hole

Levee

Distributary mouth bars
Delta front slope
Distal delta front

Prodelta

C TYPICAL PETROPHYSICAL WELL LOG

Spontaneous potential   Resistivity

coal

B Thickness of sand

thick

channel sands

delta-front sands

10 m
20 m
30 m
40 m

30 m
20 m
10 m

30 m
20 m
10 m

30 m
20 m
10 m

thin

~10 Kilometers
SCALE

Figure 2-13 Maps and borehole log of hypothetical, idealized distributary system of fluvial-dominated delta: (A) Principal features including finger-like channel deposits that parallel river courses and protrude onto shelf. (B) Contour map of aggregate thickness of sand. (C) Petrophysical log of borehole penetrating distributary channel and delta-front sands. Modified from Fisher (1969).

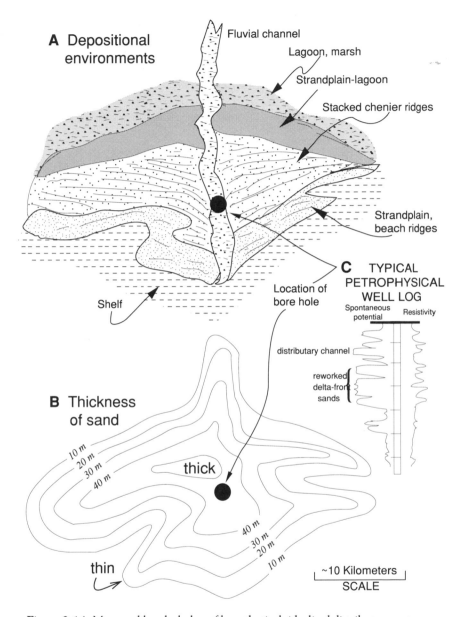

A Depositional environments

Fluvial channel

Lagoon, marsh

Strandplain-lagoon

Stacked chenier ridges

Strandplain, beach ridges

Shelf

Location of bore hole

C TYPICAL PETROPHYSICAL WELL LOG

Spontaneous potential   Resistivity

distributary channel

reworked delta-front sands

B Thickness of sand

10 m
20 m
30 m
40 m

thick

40 m
30 m
20 m
10 m

thin

~10 Kilometers
SCALE

Figure 2-14 Maps and borehole log of hypothetical, idealized distributary system of cuspate wave-dominated delta: (A) Principal features including chenier ridges and beach ridges. (B) Contour map of aggregate thickness of reworked sand deposits showing elongate form. (C) Petrophysical log of borehole showing distributary channel-sand deposits and thick, well-sorted sandstones characteristic of reworked deltaic sands. Modified from Fisher (1969).

common. Waves are so effective in removing sediment supplied by rivers that wave-dominated deltas rarely protrude far into receiving basins and are therefore generally smaller in area than fluvial-dominated deltas. Wave-dominated deltas often contain thicker sheet sands than those of fluvial deltas, and their channel, channel-mouth bar, and coastal barrier-bar deposits can be interpreted in petrophysical logs of boreholes. Examples of ancient wave-dominated deltas are described by Boyd and Dyer (1964), Fisher (1969), Fisher and others (1969), Weise (1980), and Tyler and Ambrose (1985).

Figure 2-12 emphasizes the relative influences of fluvial, wave, and tidal processes on deltaic deposits, and the continuum of morphologic characteristics that are diagnostic of the relative influence of processes that form them. These diagnostic features permit simulated deltas to be compared with actual deltas whose wave energy and discharge rates are known, such as the Mississippi, Danube, Ebro, Niger, Nile and São Francisco deltas which are compared with simulated deltas in Chapter 7.

Destructional phase of deltas

Destruction of deltaic deposits is accelerated by subsidence and sea-level changes, and by channel abandonment that deprives active delta lobes of new sediment. These changes allow waves to reshape the shoreline and rework deposits into well-sorted sheetsands, thereby creating potentially favorable conditions for hydrocarbon reservoirs or aquifers. The Mississippi delta (Figure 2-11A) provides examples of fluvial and wave-dominated features that are undergoing different degrees of destruction by waves. The active lobe at the mouth of the Mississippi River is richly supplied with sediment and is advancing at a rate of 4.5 m per year, whereas adjacent coastlines that are deprived of sediment are retreating at an average of 5.8 m per year (Saxena, 1976). Ancient deltas within the Mississippi delta complex, such as the Lafourche and St. Bernard deltas (Figure 2-11A), are being destroyed as subsidence exposes them to waves. The Chandeleur Islands are remnants of the St. Bernard delta that are being reworked by waves, and Grand Isle and adjacent barrier islands are reworked remnants of the LaFourche delta. Saxena's (1976) description of the destructional phase of a delta (Figure 2-15) is useful for interpreting deposits of reworked deltas, where waves have reworked and redistributed older deltaic sands into elongate barrier islands or sand bars. In cross-section, reworked deltaic sands form thin, areally extensive, well-sorted, quartz-rich sheet sands.

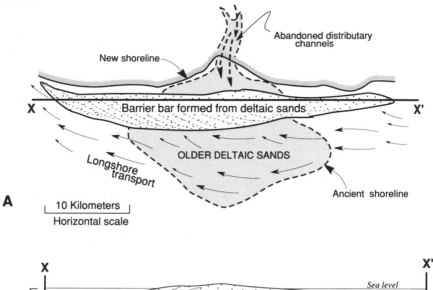

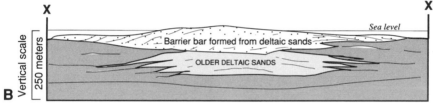

Figure 2-15 Map and longitudinal section showing older deltaic deposits that are reworked by waves: (A) Map showing outline (dashed line) of ancient delta sands formed by abandoned distributary channel being reworked to form modern elongate barrier bar. (B) Section X-X' showing barrier bar forming above older deltaic sands. Modified from Saxena (1976).

SUMMARY

This introduction to nearshore environments provides an abbreviated overview of features that WAVE and SEDSIM can mimic by representing changes in wave energy, fluvial discharge, sea level, subsidence, and other parameters that influence the development of nearshore environments. Experimental results presented in subsequent chapters can be qualitatively compared to examples here, while more quantitative comparisons are provided within chapters. Experiments closely follow descriptions of computer procedures, as in Chapter 3, which introduces procedures for representing water-wave mechanics, followed by experiments involving simulation of wave refraction, shoaling, orbital motions, and longshore currents.

Hydrodynamics
of Waves and
Nearshore Currents

Computer procedures that represent sediment transport by waves must first represent the oscillatory, orbital, steady, and turbulent motions of waves that suspend, sort, and transport sediment. This chapter describes equations used to represent these processes, and while most are used in WAVE, some are presented for background.

The receiving basin in WAVE is an initially motionless volume of water to which energy added in the form of propagating waves is subsequently dissipated as waves break and transport sediment. Wave heights, directions and periods supplied as inputs to WAVE determine the wave energy represented in an experiment, and the continuity and momentum equations ensure that mass, energy and momentum are conserved as waves move toward shore.

Adaptations of equations for representing waves involves seven main steps described in this chapter:

(1) Select waveforms (Airy and Stokes waves) to be represented in WAVE.
(2) Select equations for two-dimensional representation of shoaling waves.
(3) Select equations for wave refraction, thus adding the third dimension.
(4) Adapt a continuity equation to conserve mass.
(5) Adapt equations for conservation of momentum.
(6) Devise finite-difference procedures for solving equations numerically.
(7) Incorporate WAVECIRC as a subprogram within WAVE for solution of equations that represent mechanics of water waves.

CLASSIFICATION OF WAVES

Waveforms are divided into three main types that include (1) small-amplitude Airy and Stokes waves, (2) shallow-water cnoidal waves, and (3) solitary waves. These wave types are defined in Figure 3-1, which plots ratios of wave height to depth, and depth to wavelength, and illustrates conditions where the different types occur. Figure 3-2 defines the principal terms and symbols used to describe water waves. Theories and descriptions of the hydrodynamics of waves are provided ·by Airy (1845), Stokes (1847), Munk (1949), Morrison and Crooke (1953), Inman and Nasu (1956), Wiegel (1964), Phillips (1966), and the U.S. Army Coastal Engineering Research Center (1977).

Wave motions in deep water are generally symmetric and sinusoidal and are similar in form to Airy and Stokes waves, but waves in shallower water tend to have steeper crests and wider troughs, and are more similar to cnoidal and solitary waves. Computer procedures for representing cnoidal and solitary waves are difficult to employ (Komar 1976; Sleath, 1984; Hardy and Kraus, 1988; and Lakhan 1989), so WAVE's procedures are confined to equations that approximate Airy waves in deep water and Stokes waves in shallower water.

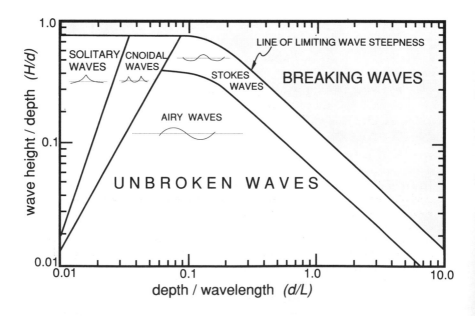

Figure 3-1 Classification of wave types based on log-log plot of wave height divided by depth (vertical axis), verus depth divided by wavelength (horizontal axis). Waves break when ratios exceed line of limiting wave steepness. After Komar (1976).

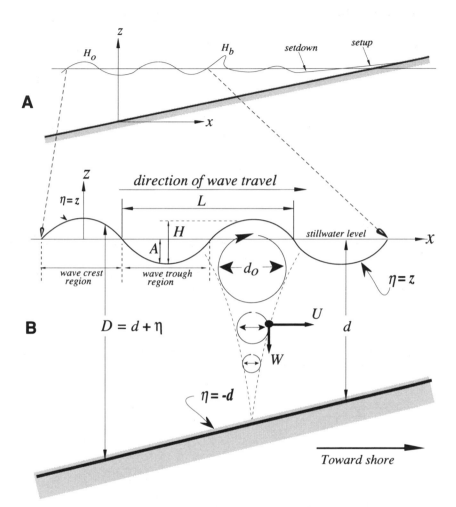

Figure 3-2 Cross section through hypothetical beach showing terms and symbols used to describe water waves: (A) Propagating ocean wave with deep-water wave height H_o, breaking-wave height H_b, and setup and setdown. (B) Enlargement showing elevation of wave surface η, wave amplitude A, wave height H, wavelength L, orbital diameter d_o, water depth d, and vertical and horizontal components of wave-induced currents W and U that correspond to coordinates x and z, respectively.

TWO-DIMENSIONAL REPRESENTATION OF SHOALING WAVES

Waves slow, shorten, steepen and refract as they travel from deep water into shallow water (Figures 3-3 and 3-4). As a wave encounters progressively shallow depths, its period does not change, but its wavelength decreases and amplitude increases to compensate for decreasing depth. Airy theory defines the displacement of a wave's upper surface η with respect to the z axis (Figure 3-2) with a harmonic function:

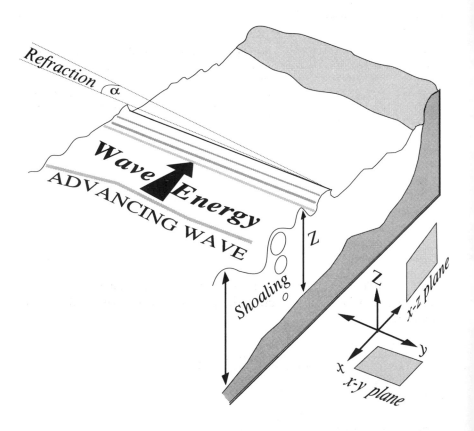

Figure 3-3   Diagram showing shoaling and refraction of waves. Equations for shoaling waves are formulated in vertical x-z plane, whereas refraction equations are formulated in horizontal x-y plane. Angle of incidence α is acute angle formed between wave crests and trend of submerged topography, which is roughly parallel to shore.

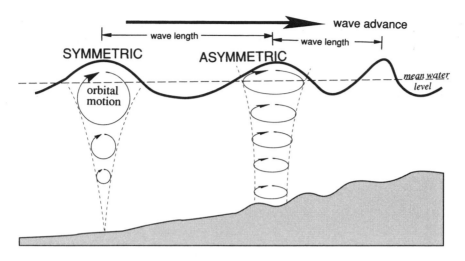

Figure 3-4 Cross section through hypothetical beach showing ellipses that represent motions of waves. As waves approach shore, wavelength decreases and wave height increases, producing waveforms that become increasingly asymmetric. In shallow water, elliptical orbitals are transformed into horizontal components that move sediment on bottom, whereas in deeper water orbitals do not affect bottom. Modified from U.S. Army Coastal Engineering Research Center (1977).

$$\eta = A\cos\theta \qquad (3\text{-}1)$$

where

    A = amplitude ($A = H/2$, where H is wave height)
    θ = phase angle in radians

In two-dimensions, phase angle θ describes the location of water particles in the x axis (Figure 3-5) at time t during the passing of waves with period T and wavelength L:

$$\eta = A\cos\left(\frac{2\pi x}{L} - \frac{2\pi t}{T}\right) \qquad (3\text{-}2)$$

where

    L = wavelength
    T = period

41

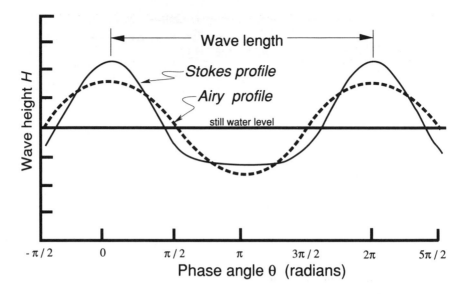

Figure 3-5   Profiles contrasting Stokes and Airy waves. Stokes profile has steeper crests and wider troughs, whereas Airy profile is perfectly sinusoidal. Phase angle θ (Equation 3-35) is function of time, wave height, length, and period. After U.S. Army Coastal Engineering Research Center (1977).

For convenience, $2\pi/L$ and $2\pi/T$ are represented by operators called wave number k and wave angular frequency ω, respectively, where 2π represents a complete circular orbit:

$$k = \frac{2\pi}{L} \tag{3-3}$$

$$\omega = \frac{2\pi}{T} \tag{3-4}$$

Substituting Equations 3-3 and 3-4 in Equation 3-2, we obtain:

$$\eta = A\cos(kx - \omega t) \tag{3-5}$$

where k is a function of wavelength and ω is a function of wave period. Holding distance x constant, Equations 3-1 to 3-5 help predict the surface locations of water particles with respect to the vertical axis z (Figures 3-2 and 3-5) as waves pass, but a more thorough description of wave hydrodynamics is required to represent the velocity and energy of waves as they move towards shore.

A wave travels a distance equal to its wavelength during one wave period. The speed at which a wave propagates is its phase velocity, or celerity C:

$$C = \frac{L}{T}$$

(3-6)

From Airy wave theory (Wiegel, 1964), wavelength can be expressed in terms of water depth:

$$L = \frac{gT^2}{2\pi} \tanh\left(\frac{2\pi d}{L}\right)$$

(3-7)

where

g = acceleration due to gravity
d = water depth
tanh = hyperbolic tangent

Combining Equations 3-6 and 3-7 gives celerity as a function of wavelength, period, and depth:

$$C = \frac{gT}{2\pi} \tanh\left(\frac{2\pi d}{L}\right)$$

(3-8)

With Equation 3-6, wave period T in Equation 3-8 can be expressed as a function of wave celerity and wave length L, so that Equation 3-8 becomes:

$$C = \left\{\frac{gL}{2\pi} \tanh\left(\frac{2\pi d}{L}\right)\right\}^{1/2}$$

(3-9)

Or, using Equation 3-3, Equation 3-9 is commonly given as a function of wave number k:

$$C = \left\{\frac{g}{k} \tanh\left(\frac{2\pi d}{L}\right)\right\}^{1/2}$$

(3-10)

Equations 3-8, 3-9, and 3-10, while redundant, are presented because derivations of other equations generally proceed from one of these three equations.

Expressions for Airy waves are simplified in deep and shallow water. In deep water where water depth d is much greater than wavelength L, $\tanh(2\pi d/L)$ approaches 1.0, and Equation 3-8 becomes:

$$C = \frac{gT}{2\pi} \tag{3-11}$$

Using Equation 3-6, Equation 3-11 can be rewritten as:

$$C_{deep} = \left(\frac{gL}{2\pi}\right)^{1/2} \tag{3-12}$$

In shallow water, $\tanh(2\pi d/L)$ approaches $2\pi d/L$ and Equation 3-8 becomes:

$$C_{shallow} = \frac{gT}{2\pi}\left(\frac{2\pi d}{L}\right) \tag{3-13}$$

Using Equation 3-6, Equation 3-13 can be rewritten as:

$$C_{shallow} = \left(gd\right)^{1/2} \tag{3-14}$$

Similarly, Equation 3-7 can be simplified in deep and shallow water:

$$L_{deep} = \frac{gT^2}{2\pi} \tag{3-15}$$

$$L_{shallow} = \left(gdT^2\right)^{1/2} \tag{3-16}$$

Substitution of Equation 3-16 in 3-2 provides a description of the displacement of the surface η of a water wave with respect to the z axis:

$$\eta = A\cos\left(\frac{2\pi x}{\left(gdT^2\right)^{1/2}} - \frac{2\pi t}{T}\right) \tag{3-17}$$

Equation 3-17 shows that as waves encounter progressively shallower depths and wave period remains constant, wavelength must decrease, and amplitude must increase if energy is to be conserved. Similarly, Equation 3-18 (Komar, 1976, p. 104) is a useful expression that relates shoaling wave height H to deep-water wave height H_0, wavelength L, and depth d :

$$H = H_0 \left(\frac{1}{\tanh(kd)\left[1 + \dfrac{2kd}{\sinh(2kd)}\right]} \right)^{1/2} \qquad (3\text{-}18)$$

where

H_0 = wave height in deep water
H = shoaling wave height

Orbital motions in oscillatory waves

Orbital motions, or "orbitals" are created by to-and-fro oscillatory movements of waves (Figure 3-4). Airy waves (Equation 3-1) in Figures 3-1, 3-2B, and 3-5 represent sinusoidal waveforms whose orbital motions are symmetric. Components of orbital motions are given by Morrison and Crooke (1953):

$$U = \frac{\pi H}{T} \frac{\cosh(2\pi\,\eta/L)}{\sinh(2\pi d/L)} \cos 2\pi \left(\frac{x}{L} - \frac{t}{T} \right) \qquad (3\text{-}19)$$

$$W = \frac{\pi H}{T} \frac{\sinh(2\pi\eta/L)}{\sinh(2\pi d/L)} \sin 2\pi \left(\frac{x}{L} - \frac{t}{T} \right) \qquad (3\text{-}20)$$

where

U and W = components of orbital velocity in x and z directions, respectively (Figure 3-2)

Orbitals in deep water are circular and do not extend to the sea bottom, but as waves enter shallow water, orbitals become increasingly elliptical and touch the bottom, creating bottom velocities capable of moving sediment (Wiegel, 1964; Komar, 1976). Velocity vectors within orbitals resolve into two horizontal velocity components, one representing motion toward shore (the propagation direction) and the other away from shore. Shoreward velocity components occur under wave

45

peaks, while seaward velocities occur under wave troughs (Figure 3-6). Bottom velocities under wave peaks are equal in magnitude, but opposite in direction to those under wave troughs.

While equations for Airy waves are widely applied (Figure 3-1) to waveforms in deep water, asymmetric waves in shallow water are more similar to Stokes waves (Figure 3-5). Equations for representing Stokes waves (Equation 3-21) include additional terms, which when added to the basic Airy sine wave in Equation 3-1, produce waveforms that have steeper peaks separated by flatter troughs (U.S.Army Coastal Engineering Research Center, 1977, p. 2-37 and 2-40), as provided by Equation 3-21:

$$\eta = \frac{H}{2}\cos\theta + \frac{\pi H^2}{8L}\frac{\cosh(2\pi d/L)}{[\sinh(2\pi d/L)]^3}\Big[2 + \cosh(4\pi d/L)\Big]\cos 2\theta \qquad (3\text{-}21)$$

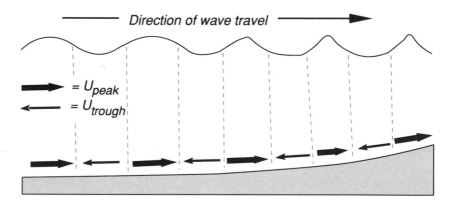

Figure 3-6 Resolved bottom velocities beneath waves differ in magnitude and direction. Bold arrows represent shoreward velocities U_{peak} that are larger than return velocities U_{trough}. Adapted from Clifton and Dingler (1984).

Figure 3-5 compares profiles of Airy and Stokes waves. Stokes theory is widely applied in shallow water because it predicts waveforms that conform closely to those observed in shallow water where waves develop progressively steeper crests and wider, flatter troughs as they move toward progressively shallower water.

Morrison and Crooke (1953) and Inman and Nasu (1956) provide approximations derived from Stokes theory for orbital velocities U and W in the x and z directions (Figure 3-2) :

$$U = \frac{\pi H}{T}\frac{\cosh(2\pi\eta/L)}{\sinh(2\pi d/L)}\cos 2\pi\left(\frac{x}{L}-\frac{t}{T}\right)+$$
$$\frac{3}{4}\frac{\pi^2 H^2}{LT}\frac{\cosh(4\pi\eta/L)}{[\sinh(2\pi d/L)]^4}\cos 4\pi\left(\frac{x}{L}-\frac{t}{T}\right)$$

(3-22)

$$W = \frac{\pi H}{T}\frac{\sinh(2\pi\eta/L)}{\sinh(2\pi d/L)}\sin 2\pi\left(\frac{x}{L}-\frac{t}{T}\right)+$$
$$\frac{3}{4}\frac{\pi^2 H^2}{LT}\frac{\sinh(4\pi\eta/L)}{[\sinh(2\pi d/L)]^4}\sin 4\pi\left(\frac{x}{L}-\frac{t}{T}\right)$$

(3-23)

Under wave peaks and troughs, where θ is either 0, π, or 2π (Figure 3-5), Equations 3-22 and 3-23 can be simplified, and expressions for horizontal water-particle velocities below wave peaks U_{peak} and troughs U_{trough} become:

$$U_{peak} = \frac{\pi H}{T}\frac{\cosh(2\pi\eta/L)}{\sinh(2\pi d/L)}+\frac{3}{4}\frac{\pi^2 H^2}{LT}\frac{\cosh(4\pi\eta/L)}{[\sinh(2\pi d/L)]^4}$$

(3-24)

$$U_{trough} = -\frac{\pi H}{T}\frac{\cosh(2\pi\eta/L)}{\sinh(2\pi d/L)}+\frac{3}{4}\frac{\pi^2 H^2}{LT}\frac{\cosh(4\pi\eta/L)}{[\sinh(2\pi d/L)]^4}$$

(3-25)

Equations 3-24 and 3-25 and Figure 3-6 show that horizontal bottom currents below wave troughs U_{trough} are opposite in direction to currents below wave peaks U_{peak}. U_{peak} currents move toward shore and are positive by convention, whereas U_{trough} currents move offshore.

The second term in Equations 3-24 and 3-25 deals with the difference in magnitude between shoreward and seaward velocities. Steeper crests and wider troughs of Stokes waves produce asymmetrical orbital motions, unlike Airy waves whose orbital motions are symmetrical. Orbital motions under wave crests are stronger and of shorter duration than orbital motions under wave troughs, creating horizontal velocities beneath wave peaks that are greater (large arrows in Figure 3-6) than those below wave troughs (small arrows). The differing velocities predicted by Stokes wave theory are important because the differences between offshore velocities and onshore velocities on the bottom are capable of sorting and transporting sediment, and may have a large influence on directions of sand transport.

Wave-induced shear stress

Orbital motions of waves increase shear stress at the sea bottom, allowing waves to suspend sediment that can then be moved by other currents. Thus, waves exert an additional shear in addition to shear stress produced by other types of flow. The maximum orbital velocity U_{max} helps in estimating wave-induced bottom shear stress. The maximum orbital velocity occurs as a wave peak passes, where $\theta=2\pi$, permitting Equation 3-24 to be simplified:

$$U_{max} = \frac{\pi H}{T \sinh(2\pi d/L)} \tag{3-26}$$

At the ocean bottom, orbital motions are ellipsoid or nearly horizontal, and their long axes (Figure 3-2) are given by:

$$d_{o_{max}} = U_{max}\left(\frac{T}{\pi}\right) \tag{3-27}$$

where

$d_{o_{max}}$ = length of the long axis of ellipse near bottom (Figure 3-2)

Substituting Equation 3-27 into 3-26 gives a standard formulation of the maximum orbital diameter at the sea bottom, which is important for some calculations involving bottom shear stress imparted by waves:

$$d_{o_{max}} = \frac{H}{\sinh(2\pi d/L)} \tag{3-28}$$

Breaking waves and heights of breaking waves

The amount of energy released by breaking waves can be estimated from the heights of waves at the moment that they break. Wiegel (1964), Weggel (1972), U.S. Army Corps of Engineers Research Center (1977), and Douglass and Weggel (1988) show that the depth where waves break can be predicted from water depth and either wave height or wavelength. The U.S. Army Corps of Engineers Research Center (1977) provides a ratio for predicting where waves break as a function of wave height and water depth:

$$\frac{H_b}{d_b} = 0.78 \tag{3-29}$$

where

H_b = breaking-wave height
d_b = depth at which wave breaks

A more precise formulation is given by Le Mahaute (1961):

$$\frac{H_b}{L_b} = 0.12 \tanh 2\pi \left(\frac{d_b}{L_b} \right) \tag{3-30}$$

where

L_b = wavelength at breaker zone, generally expressed as $2\pi/k$
         (from Equation 3-3)

REPRESENTATION OF REFRACTING WAVES IN THE THIRD DIMENSION

Figure 3-2 defines characteristics of shoaling waves in two dimensions, along the x and z axes. Equations 3-1 through 3-21, derived from Airy and Stokes wave theories, describe how waves slow, shorten, and steepen as they travel from deep water into shallow water. These "two-dimensional" equations are used to determine wave height H and wavelength L, permitting bottom velocities (Equations 3-24 and 3-25) to be determined. A third dimension is added to the representation of shoaling waves by incorporating refraction of waves in the x-y plane (Figure 3-3). Refraction in the x-y plane involves additional equations for describing changes in wave angles and heights as waves move toward shore.

Wave angle
    Waves often approach the shore at an oblique angle, causing paths of waves to bend or refract because the inshore part of the wave front moves slower than the wave front in deep water. Thus, the refraction of water waves is analogous to refraction of light rays. The refraction of wave orthogonals in Figures 3-7 and 3-8 is related to change in phase velocity of advancing waves as they move over changing depths. A first approximation of wave refraction as waves move from deep to shallow water over a uniformly dipping slope is provided by Snell's law:

$$\frac{\sin \beta_d}{C_d} = \frac{\sin \beta_s}{C_s} \tag{3-31}$$

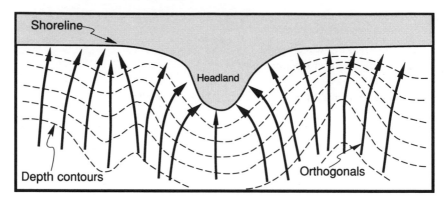

Figure 3-7   Map showing wave orthogonals where submerged topography is irregular, causing orthogonals to converge in shallow water and diverge in deeper water. Wave-induced currents increase where wave orthogonals converge, and decrease where orthogonals diverge. Scale can vary greatly. Modified from U.S. Army Coastal Engineering Research Center (CERC), 1977.

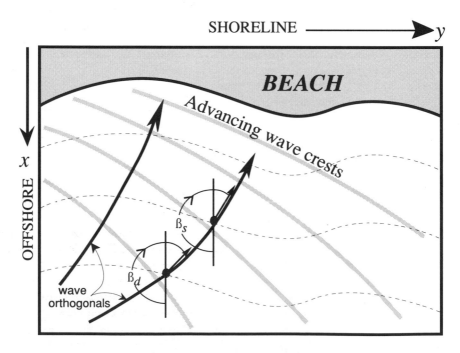

Figure 3-8   Map of coastline showing refraction of advancing waves approaching shallow water. Dashed contours represent depth. Wave angle ß is measured clockwise from x axis, with $ß_d$ in deep water and $ß_s$ in shallow water.

where

β_d = wave angle in deep water (Figure 3-8)
β_s = wave angle in shallow water (Figure 3-8)
C_d = phase velocity in deep water (Equation 3-12)
C_s = phase velocity in shallow water (Equation 3-14)

Wave height
    Because wave height is inversely proportional to phase velocity, Equation 3-32 describes effects of refraction on shoaling wave heights, predicting that wave height will increase as wave angle β (Figure 3-8) decreases:

$$H_o \sin\beta_d \approx H_s \sin\beta_s \qquad (3\text{-}32)$$

where

H_o = wave height in deep water
H_s = wave height in shallow water

Noda and others (1974) utilize this general relationship to predict shoaling wave heights H:

$$H = H_o \frac{1}{\sqrt{\beta}} \left(\frac{1}{\tanh(kd)\left[1 + \dfrac{2kd}{\sinh(2kd)}\right]} \right)^{1/2} \qquad (3\text{-}33)$$

where

H_o = wave height in deep water
β = $\cos\beta_s / \cos\beta_d$ as defined in Figure 3-8

$$\frac{1}{\sqrt{\beta}} = \left(\cos\beta_d / \cos\beta_s\right)^{1/2} = \text{``refraction coefficient''} \qquad (3\text{-}34)$$

The three-dimensional derivation of the equations of Noda and others given by Equation 3-33 is nearly identical to Equation 3-18, except that Equation 3-33 includes a "refraction coefficient" allowing refraction to be included in the description of shoaling waves.

Wavelength and wave number

In three dimensions, Noda and others (1974) showed that it is more convenient to solve for wave number k (Equation 3-3) than to solve for wavelength L. Equation 3-1 can be further expanded to yield an expression that predicts the wave number in three dimensions. Equation 3-1 defines the sinusoidal motion of a propagating wave over one wave period (from 0 to 2π), and inspection of Equations 3-1 and 3-2 shows that phase angle θ is a function of wavelength and wave period given by:

$$\theta = \frac{2\pi}{L} x - \frac{2\pi}{T} t \qquad (3\text{-}35)$$

Terms for wave number k (Equation 3-3) and wave angular frequency ω (Equation 3-4) are embodied in Equation 3-35, so Equations 3-1, 3-2, 3-3, and 3-4 can be combined with Equation 3-35 to provide an alternative expression for Equation 3-1:

$$\eta = \frac{H}{2} \cos (kx - \omega t) \qquad (3\text{-}36)$$

Therefore, a new phase function ϕ incorporating the horizontal dimension y can be defined:

$$\phi = f[k(x, y) - \omega(t)] \qquad (3\text{-}37)$$

Equations 3-1 and 3-37 can be rewritten as a three-dimensional expression describing the elevation of wave surface η with respect to spatial directions z, x, and y, and time t:

$$\eta_{(x,y,t)} = A_{(x,y,t)} \cos \phi_{(x,y,t)} \qquad (3\text{-}38)$$

where

η = surface displacement along z-axis (Figure 3-2)
x,y = horizontal coordinates (Figures 3-2 and 3-3)
ϕ = phase function in radians
A = wave amplitude (= $H/2$, where H is wave height)

As in the two-dimensional derivation for η, Equation 3-38 can be expressed in terms of wave number k and wave frequency ω. In three dimensions, the wave number is a vector quantity equal to the gradient of the scalar phase function ϕ:

$$\vec{k} = i \frac{\partial \phi}{\partial x} + j \frac{\partial \phi}{\partial y} + k \frac{\partial \phi}{\partial z} \tag{3-39}$$

where

$\vec{k}$ = wave number vector

Equation 3-39 can be simplified to:

$$\vec{k} = \nabla \phi \tag{3-40}$$

where

$$\nabla = i \frac{\partial}{\partial x} + j \frac{\partial}{\partial y} + k \frac{\partial}{\partial z} \tag{3-41}$$

i, j, k = unit vectors
x, y, z = spatial coordinates (Figure 3-3)

The scalar phase function ϕ is similar to the two dimensional phase function θ in Equation 3-35, but defines the rate of fluctuation of the ocean's surface in three dimensions. Wave frequency ω is a scalar quantity that is a function of wave phase ϕ with respect to time and can be expressed as a function of time:

$$\overline{\omega} = - \frac{\partial \phi}{\partial t} \tag{3-42}$$

where

$\overline{\omega}$ = wave frequency, where bar denotes time-averaging over one wave period

Noda and others (1974) and Ebersole and Dalrymple (1979) combine Equations 3-40 and 3-42 to derive the equation for describing the "conservation of waves:"

$$\frac{\partial \vec{k}}{\partial t} + \nabla \overline{\omega} = 0 \tag{3-43}$$

Equation 3-43 predicts that a change in wave number with respect to time must equal the gradient of wave frequency if the wave is irrotational. Thus, wave number k can be determined from Equation 3-43, where wave frequency ω is equal to $2\pi/T$ and wave period T is constant. Where other wave-induced currents are present, Ebersole and Dalrymple (1979) show that wave number is a function:

$$f(k) = \left[gk \ \tanh(kd)\right]^{1/2} + uk \ \cos\theta + vk \ \sin\theta - \frac{2\pi}{T} \qquad (3\text{-}44)$$

This relationship is a convenient formulation for predicting wave number that embodies the shoaling nature of waves in three dimensions.

CONTINUITY EQUATIONS

Longuet-Higgins (1970) argues that momentum of incoming waves drives nearshore current systems and that momentum must be conserved in equations that represent nearshore systems. WAVE incorporates the ideas of Longuet-Higgins by representing waves using equations that satisfy the conservation laws, and allowing the momentum of propagating waves to create nearshore currents. However, equations used by WAVE are formulated to provide time-averaged and depth-averaged values to simplify computer procedures for representing wave-induced currents. Equations used by WAVE provide values at each grid cell that are depth-averaged because they represent an average midway between the bottom and the upper surface over total depth D. Similarly equations used by WAVE provide time-averaged values in grid cells, where values are averaged over one wave period T. Thus, time-averaged and depth-averaged values provide a quasi three-dimensional representation of wave-induced currents, but simplify the continuity and momentum equations that follow. Additional descriptions of continuity and momentum equations used in WAVE are provided by Longuet-Higgins (1970), Noda and others (1974), and Ebersole and Dalrymple (1979).

Boundary conditions for free-flow surface flow

Kinematic boundary conditions are employed in the continuity and momentum equations that describe waves. The interface between water and air is a "free" surface that continually changes in space and time (Figure 3-9). By contrast, the surface at the sea bottom is "rigid" because changes in elevation by erosion or deposition are small relative to changes in wave heights. A function f in Equation 3-45 describes a finite fluctuating surface that forms a boundary which water particles in a wave cannot cross:

$$f(x, y, z, t) = 0 \qquad (3\text{-}45)$$

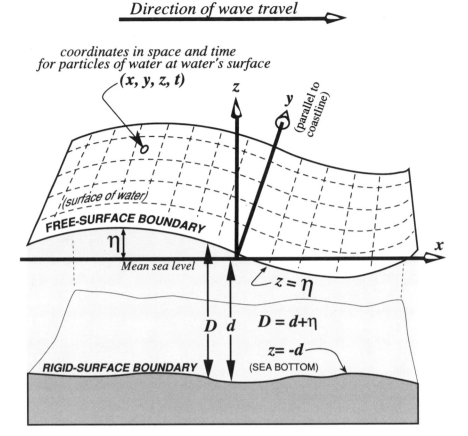

Figure 3-9 Perspective diagram defining free-surface and rigid-surface boundaries of water waves. Axes *x*, *y*, and *z* define coordinates. Water's upper surface η is free-surface boundary represented by coordinates *x*, *y*, and *z* and time *t*. Rigid surface boundary is defined by sea bottom at depth *d* measured from mean sea level. Water particles cannot cross boundaries.

where

f = function defining kinematic boundary surface

t = time

x, y, z = spatial coordinates (Figure 3-9)

Figure 3-9 demonstrates that z is equal to η at the free-surface boundary (surface of the water wave), where η is the instantaneous height of the water surface. Thus, the kinematic boundary condition at water's surface is:

$$f^{surface}(x, y, z, t) = z - \eta(x, y, t) = 0 \tag{3-46}$$

At the sea bottom z is equal to $-d$, and the rigid boundary surface is:

$$f_{bottom}(x, y, z, t) = z + d(x, y, t) = 0 \tag{3-47}$$

where

z = surface elevation of wave with respect to z axis (Figures 3-2 and 3-9)
d = water depth with respect to z axis (Figures 3-2 and 3-9)

Equation 3-45 can be expressed in terms of velocity components of water particles confined to the boundary surface, allowing the kinematic boundary surface to be expressed as a function of water particle velocities, local grid coordinates, time, and three-dimensional space. The general kinematic equation describing the fluctuation of surface f in three-dimensions is:

$$\frac{\partial f}{\partial t} + u \frac{\partial f}{\partial x} + v \frac{\partial f}{\partial y} + w \frac{\partial f}{\partial z} = 0 \tag{3-48}$$

where

u, v, w = velocity components in x, y, z directions (Figures 3-2 and 3-9)

Thus, at the free surface where z equals η, the height of the water surface varying through time is:

$$\frac{\partial \eta}{\partial t} + u_\eta \frac{\partial \eta}{\partial x} + v_\eta \frac{\partial \eta}{\partial y} - w_\eta = 0 \tag{3-49}$$

For the bottom boundary surface, where z equals $-d$, Equation 3-49 becomes:

$$\frac{\partial \eta}{\partial t} - u_{-d} \frac{\partial \eta}{\partial x} - v_{-d} \frac{\partial \eta}{\partial y} + w_{-d} = 0 \tag{3-50}$$

Assuming the mass of individual water particles is conserved, we can write a general form of the continuity equation that is similar to Equation 3-48, except that a term for water density is introduced. Equation 3-51 is an expression for the three-dimensional continuity equation, which states that mass is conserved as a waveform propagates through water.

$$\frac{\partial \rho}{\partial t} + \frac{\partial(\rho u)}{\partial x} + \frac{\partial(\rho v)}{\partial y} + \frac{\partial(\rho w)}{\partial z} = 0 \tag{3-51}$$

where

ρ = density of water

Ebersole and Dalrymple (1979) derived Equation 3-52 by integrating Equation 3-51 over the total depth, $d + \eta$ shown in Figure 3-9. Equation 3-52 shows that w, the water velocity component in the z direction, is eliminated by integrating over the total depth from $z = \eta$ (water surface) to $z = -d$ (sea bottom).

$$\frac{\partial}{\partial t} \int_{-d}^{\eta} \rho \, dz + \frac{\partial}{\partial x} \int_{-d}^{\eta} \rho u \, dz + \frac{\partial}{\partial y} \int_{-d}^{\eta} \rho v \, dz = 0 \tag{3-52}$$

Finally, we note that u and v in Equation 3-52 are composed of three velocity components expressed in Equations 3-53 and 3-54, namely (1) a mean current, (2) a wave-induced current, and (3) a component of turbulence:

$$u = U_{mean} + U_{orb} + U_{turb} \tag{3-53}$$
$$v = V_{mean} + V_{orb} + V_{turb} \tag{3-54}$$

where

u = total velocity component in x direction
v = total velocity component in y direction
U, V_{mean} = mean currents, in x and y directions
U, V_{orb} = currents produced by orbital motions of waves in x and y directions
U, V_{turb} = eddy currents produced by turbulence in x and y directions

However, Ebersole and Dalrymple (1979) showed that integration of turbulent velocities over one wave period (time averaging) yields turbulent velocities U_{turb} and V_{turb} that are negligible, permitting Equations 3-53 and 3-54 to be substituted into Equation 3-52 to simplify the continuity equation, where U and V represent depth-averaged mean currents:

$$\frac{\partial \bar{\eta}}{\partial t} + \frac{\partial}{\partial x}(UD) + \frac{\partial}{\partial y}(VD) = 0 \tag{3-55}$$

where

$\overline{\eta}$ = water level elevation (Figures 3-2, 3-9), bar denotes time-averaging during one wave period

U = depth-averaged and time-averaged U_{mean} plus wave induced currents U_{orb} in x-coordinate direction

V = depth-averaged and time-averaged currents V_{mean} plus wave induced currents V_{orb} in y-coordinate direction

D = total depth $d + \eta$ (Figures 3-2 and 3-9)

Therefore Equation 3-55 can predict nearshore currents produced by mean currents, as well as currents produced by asymmetrical, time-varying, orbital wave motions. Furthermore, current velocities are depth-averaged, thus simplifying the three-dimensional representation by dealing only with average quantities with respect to depth.

MOMENTUM EQUATIONS FOR CONSERVATION OF ENERGY AND MOMENTUM

Waves have momentum, most of which is dissipated in the surf zone by conversion to turbulence and heat by breaking waves. Kinetic energy not converted to heat produces longshore currents, rip currents, and changes in sea level called "set up" (Figure 3-2). Representation of these processes requires equations that ensure that momentum and energy are conserved.

Longshore currents and rip currents produced by excess momentum of incoming waves are depth averaged in WAVE, so that they are expressed solely with components U and V, in the x-y plane (Figure 3-3). Ebersole and Dalrymple (1979) refer to U and V in Equation 3-55 as "mass transport velocities". Mass transport velocities may be large enough to initiate movement of sand and should be incorporated in equations for sediment transport. Mass-transport velocities are derived from two equations of motion that define a condition where momentum is conserved as waveforms propagate through water. Equations 3-56 and 3-57 are expressions for the x and y momentum equations:

$$\frac{\partial u}{\partial t} + \frac{\partial u^2}{\partial x} + \frac{\partial uv}{\partial y} + \frac{\partial uw}{\partial z} = \frac{-1}{\rho} \frac{\partial P}{\partial x} + \frac{1}{\rho}\left[\frac{\partial \tau_{xx}}{\partial x} + \frac{\partial \tau_{yx}}{\partial y} + \frac{\partial \tau_{zx}}{\partial z} \right] \qquad (3\text{-}56)$$

$$\frac{\partial v}{\partial t} + \frac{\partial uv}{\partial x} + \frac{\partial v^2}{\partial y} + \frac{\partial vw}{\partial z} = \frac{-1}{\rho} \frac{\partial P}{\partial y} + \frac{1}{\rho}\left[\frac{\partial \tau_{xy}}{\partial x} + \frac{\partial \tau_{yy}}{\partial y} + \frac{\partial \tau_{zy}}{\partial z} \right] \qquad (3\text{-}57)$$

where

P = fluid pressure

τ_{xx}, τ_{yx}, τ_{zx} = sum of local wave stresses including bottom stresses and surface stresses due to wind in the x, y, and z directions

Equations 3-56 and 3-57 can be manipulated in the same way as the continuity equation to derive Equations 3-58 and 3-59, which are time-averaged over one wave period and integrated over total water depth:

$$\frac{\partial}{\partial t}\int_{-d}^{\eta} u\,dz + \frac{\partial}{\partial x}\int_{-d}^{\eta} u^2\,dz + \frac{\partial}{\partial y}\int_{-d}^{\eta} uv\,dz = \frac{-\partial}{\rho\partial x}\overline{\int_{-d}^{\eta} P\,dz} + \frac{1}{\rho}\left\{\overline{P_{-d}\frac{\partial d}{\partial x}} + \overline{\tau}_{sx} - \overline{\tau}_{bx}\right\} \qquad (3\text{-}58)$$

$$\frac{\partial}{\partial t}\int_{-d}^{\eta} v\,dz + \frac{\partial}{\partial x}\int_{-d}^{\eta} v^2\,dz + \frac{\partial}{\partial y}\int_{-d}^{\eta} uv\,dz = \frac{-\partial}{\rho\partial y}\overline{\int_{-d}^{\eta} P\,dz} + \frac{1}{\rho}\left\{\overline{P_{-d}\frac{\partial d}{\partial y}} + \overline{\tau}_{sy} - \overline{\tau}_{by}\right\} \qquad (3\text{-}59)$$

where

P = fluid pressure

$\overline{P_{-d}}$ = time-averaged pressure at sea bottom, where bar denotes time averaging during one wave period (Figures 3-2, 3-9)

$\overline{\tau}_{sx}$, $\overline{\tau}_{sy}$ = time-averaged shear stress at water surface, due to wind, in x and y directions (Figure 3-10), bar denotes time-averaging during one wave period

$\overline{\tau}_{bx}$, $\overline{\tau}_{by}$ = time-averaged shear stress at sea bottom in x and y directions (Figure 3-10), where bar denotes time-averaging during one wave period

To derive time-averaged, depth-integrated current velocities, Ebersole and Dalrymple (1979) substituted Equations 3-53 and 3-54 into Equations 3-56 and 3-57, neglected turbulent fluctuations U_{turb}, and assumed that density ρ is constant. They also assumed that sea water is non-viscous, so that no horizontal viscous stress exists. Thus τ_{yx} and τ_{xx} are equal to zero, but surface stress due to wind τ_s and bottom stresses due to friction τ_b remain in the expressions. The mean pressure at the sea bottom P_{-d} is defined as the sum of the dynamic, or wave-induced pressure at the bottom, and hydrostatic pressure:

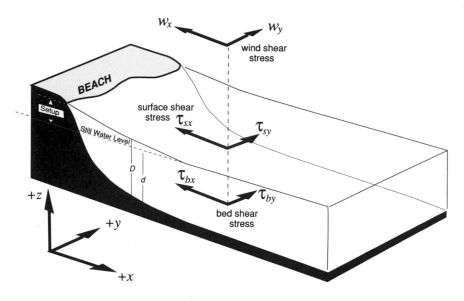

Figure 3-10 Hypothetical beach showing forces induced by wind stress w, surface shear stress τ_s, and bottom shear stress τ_b. Subscripts x and y represent coordinates where wind and shear forces are written w_x, w_y, τ_{sx}, τ_{sy}, τ_{bx}, and τ_{by}. Wave set-up involves increase in water elevation toward shore as water piles up near shore. Modified from U.S. Army Coastal Engineering Research Center (1977).

$$\overline{P_{-d}} = \overline{P_{dyn_{-d}}} + \rho g(d + \overline{\eta})$$

(3-60)

where

$\overline{P_{-d}} =$ mean, time-averaged pressure at bottom

$\overline{P_{dyn-d}} =$ mean dynamic, or wave induced pressure (time averaged)

Equation 3-60 shows that in addition to normal hydrostatic pressure, there is a dynamic or wave-induced pressure exerted as waves propagate towards shore. This wave-induced pressure is an "excess momentum flux" that contributes to the total pressure exerted by a wave and is important because it drives nearshore circulation systems. As waves advance towards shore, dynamic pressures created by orbital motions of waves create longshore, rip, and other nearshore currents. Longuet-Higgins and Stewart (1962) refer to this excess flow of momentum due to waves as "radiation stress". Longuet-Higgins (1970) derived Equations 3-61 and 3-62 as expressions for components of radiation stress (excess momentum flux) in the x and y directions:

$$S_{xx} = E \left[\frac{2kd}{\sinh(2kd)} + \frac{1}{2} \right] \qquad (3\text{-}61)$$

$$S_{yy} = E \left[\frac{kd}{\sinh(2kd)} \right] \qquad (3\text{-}62)$$

where

S_{xx} = excess momentum flux along direction of wave advance, parallel to x-axis

S_{yy} = excess momentum flux in direction of wave crests, parallel to y-axis

E = wave energy per unit length of wave crest given by Komar (1976) as:

$$E = \frac{1}{8} \rho g H^2 \qquad (3\text{-}63)$$

Noda and others (1974) incorporate terms for radiation stress in the general momentum equations expressed by Equations 3-56 and 3-57. Equations 3-64 and 3-65 are simplified expressions of momentum equations derived by Ebersole and Dalrymple (1979):

$$\frac{\partial U}{\partial t} = -g \frac{\partial \overline{\eta}}{\partial x} - \frac{1}{\rho(d+\overline{\eta})} \left[\frac{\partial S_{xx}}{\partial x} + \frac{\partial S_{xy}}{\partial y} + \overline{\tau}_{sx} - \overline{\tau}_{bx} \right] \qquad (3\text{-}64)$$

$$\frac{\partial V}{\partial t} = -g \frac{\partial \overline{\eta}}{\partial y} - \frac{1}{\rho(d+\overline{\eta})} \left[\frac{\partial S_{yx}}{\partial x} + \frac{\partial S_{yy}}{\partial y} + \overline{\tau}_{sy} - \overline{\tau}_{by} \right] \qquad (3\text{-}65)$$

where

$$S_{xy} = S_{yx} = \frac{E}{2} n \sin 2\theta \qquad (3\text{-}66)$$

$$n = \frac{C_g}{C} = \frac{1}{2}\left[1 + \frac{2kd}{\sinh(2kd)}\right] \tag{3-67}$$

$$C_g = \frac{C}{2}\left(1 + \frac{2kd}{\sinh(2kd)}\right) \tag{3-68}$$

and:

- U = depth-averaged mean currents U_{mean} and wave-induced currents U_{orb} in x direction
- V = depth-averaged mean currents V_{mean} and wave-induced currents V_{orb} in y direction
- C_g = group velocity or velocity of energy propagation (Longuet-Higgins, 1970)
- C = celerity or phase velocity (Equations 3-8, 3-9, and 3-10)

Equations 3-64 and 3-65 provide a means of directly predicting mass transport velocities U and V. Furthermore, radiation stresses are included in the momentum equations, which when incorporated, allow WAVE to simulate longshore and rip currents, as well as onshore-offshore currents produced by oscillatory motions of waves. Bottom friction and surface stresses created by wind are also included in Equations 3-64 and 3-65.

Radiation stresses

Longuet-Higgins' (1970) equations for radiation stress (Equations 3-61, 3-62, and 3-66) suggest that longshore and rip currents are dominant inside the breaker zone and observations on modern beaches support his conclusions (J. R. Dingler and J. C. Ingle, personal communication). Radiation stresses are concentrated within the breaker zone, where the dissipation of wave energy is greatest, causing longshore velocities to be strongest near the breaker line while decreasing toward shore. However, studies of beaches show that longshore velocities do not decrease linearly toward shore. Turbulence induced by breaking waves causes lateral mixing in which water particles adjacent to breaking waves are set in motion, transferring momentum within the breaker zone.

Calculation of radiation stresses without effects of lateral mixing are unrealistic in that a linear velocity profile is produced in which longshore velocities are zero seaward of the breaker zone, greatest at the breaker zone, and decrease linearly to zero toward the shore (Figure 3-11). When effects of lateral mixing are included, a more realistic geographic distribution of velocities within the surf zone is obtained. WAVE deals with the distribution of longshore velocities by using equations of Ebersole and Dalrymple (1979) that yield a nonlinear distribution produced by radiation stresses and lateral mixing.

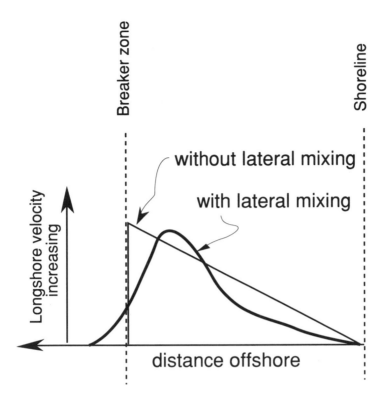

Figure 3-11 Plot perpendicular to shore comparing computed longshore velocities with respect to distance from shore. Curve represents velocities that include effects of lateral mixing, and is more realistic than straight-line plot computed without lateral mixing. From Longuet-Higgins (1970).

Surface shear stress

Surface shear stress (Figure 3-10) is caused entirely by wind and is important in the solution of Equations 3-64 and 3-65. Van Dorn (1953) and Birkemeier and Dalrymple (1975) provide simple equations to calculate the x and y components of shear stress caused by wind acting on the water's surface, where variables are schematically defined in Figure 3-10.

$$\overline{\tau}_{sx} = \rho K |w| w_x \tag{3-69}$$

$$\overline{\tau}_{sy} = \rho K |w| w_y \tag{3-70}$$

where

$$\overline{\tau}_{sx} \; = \; \text{time-averaged wind stress in } x \text{ direction}$$

$$\overline{\tau}_{sy} \; = \; \text{time-averaged wind stress in } y \text{ direction}$$

w = magnitude of wind speed
w_x = x component of wind speed
w_y = y component of wind speed
ρ = water density
K = stress coefficient

Van Dorn proposed Equations 3-71 and 3-72 that relate wind stress coefficient K to wind speed w :

$$K = K_1 \; \text{ for } w \leq \; w_c \tag{3-71}$$

$$K = K_1 + K_2 \left(1 - \frac{w_c}{w} \right)^2 \text{ for } w > w_c \tag{3-72}$$

where

w_c = critical wind speed = 14 knots (one knot = 0.5144 meters/sec)
K_1 = 1.1 x 10^{-6} (dimensionless)
K_2 = 2.5 x 10^{-6} (dimensionless)

Bottom shear stress

Bottom shear stress is used in Equations 3-64 and 3-65 for calculating wave-induced currents, and is also important in equations for sediment transport. Bagnold (1963) provides a general equation for fluid stress τ_f imparted by steady fluid flow:

$$\tau_f = C_d \, v^2 \tag{3-73}$$

where

C_d = coefficient of friction, or drag
v = flow velocity

However, waves induce an additional oscillatory component of flow, where currents have both a steady and an oscillatory component. Birkemeier and Dalrymple (1975) and Ebersole and Dalrymple (1979) provide an equation for time-averaged bottom shear stress τ_b, based on Equation 3-73.

$$\overline{\tau}_b = \frac{1}{8}\rho\, f\vec{U}_t \bullet \left|\vec{U}_t\right| \tag{3-74}$$

where

U_t = total velocity vector (denoted by arrow), including mean and orbital currents (may be positive or negative)

f = wave-induced friction factor, typically 0.01-0.08 (Jonsson, 1966)

Jonsson (1966), Komar and Miller (1973), Madsen and Grant (1976), Dingler (1979) and Hattori (1982) discuss usage and derivation of similar equations for bed shear stress. An equation for the critical shear stress required to initiate sediment transport may take a form similar to Equation 3-74, where maximum orbital velocity U_{max} (Equation 3-26) replaces U_t in Equation 3-74.

COMPUTER PROCEDURES FOR SIMULATING WAVES

Numerous computer procedures have been proposed to represent properties of shoaling waves. Early programs by Wilson (1966), and Dobson (1967) used ray tracing to calculate shoaling properties of waves, where ray orthogonals perpendicular to wave crests are traced for waves as they bend or refract over irregular submerged surfaces. Ray-tracing procedures have major disadvantages, however, in that calculations are provided only along rays, and that grids of wave heights and wave directions are not generated, thus requiring interpolation to provide them in a grid format.

WAVE, like most computer models, employs regular grids that need information at every grid point, so schemes that represent wave parameters within a fixed grid scheme are desirable. Finite-difference solutions allow wave properties such as wave height and wave direction to be calculated over a uniform grid, and are superior to ray-tracing methods because interpolation is not required to provide wave characteristics at each grid node. Noda and others (1974) developed a finite-difference procedure for simulating wave refraction and prediction of circulation patterns of wave-induced currents. Liu and Mei (1974) extended their procedure with a program that simulates effects of wave-induced setup. Birkemeier and Dalrymple (1975) used algorithms developed by Noda and others and by Liu and Mei to develop a more rigorously time-dependent procedure in which calculations are time averaged over a single wave period. Ebersole and Dalrymple (1979) in turn extended this work by developing a more complete time-dependent computing procedure that simulates wave refraction, wave-induced nearshore circulation patterns, wave-current interactions, wave setup, convective accelerations, lateral mixing, radiation stress, and wind effects. In summary, WAVE includes finite-difference procedures developed by Ebersole and Dalrymple (1979), which are

65

based on work by Noda and others (1974), Liu and Mei (1974), and Birkemeier and Dalrymple (1975).

Horikawa (1988) and Lakhan (1989) review other procedures for simulating hydrodynamics of waves. Kurihara (1965), Bowen (1969), Ito and Katsutoshi (1972), Noda (1972), Bruno and Gable (1976), Fox and Davis (1979), Felder and Fisher (1980), Ozasa and Brampton (1980), Lakhan (1982), Vemulakonda and others (1982), Birkemeier (1984), Boer and others (1984), Davies (1987), De Vriend (1987), Hardy and Kraus (1988), and Thornton and Guza (1989) also provide diverse computer programs for simulating waves.

WAVE's computer programs for representing waves

WAVE represents properties of shoaling waves with several smaller computer sub-programs, which together form module WAVECIRC adapted from a program by Ebersole and Dalrymple (1979). Martinez (1987b) and Martinez and Harbaugh (1989) review WAVECIRC's organization, calibration, use, performance, and sensitivity to various input parameters. However, computer procedures from other sources could be substituted for WAVECIRC. Our goal is to outline subprograms for linking specific computer procedures that can work together to simulate wave systems. As computer procedures improve for representing wave hydrodynamics, their incorporation in WAVECIRC is encouraged.

Numerical methods for solving equations

WAVE employs differential equations that govern wave refraction, wave-induced nearshore circulation patterns, wave-current interaction, wave setup, convective accelerations, lateral mixing, and wind effects, that are transformed to finite-difference approximations for numerical solutions to solve the continuity equation and other sets of equations that describe wave motion. These equations are time averaged over a single wave period and integrated over total depth at each grid location. Equations 3-51 3-56, and 3-57 that describe conservation of mass and momentum and the shoaling characteristics of waves cannot be solved analytically, but can be solved numerically. Numerical solutions utilize finite-difference approximations of wave properties at each cell within a uniform grid. Each cell has a specific depth during each time increment, and numerical solutions for wave number k and other wave parameters describing shoaling waves are calculated from values from preceding time steps or from values in adjacent grid cells.

Conventions defining a cellular grid are shown in Figure 3-12. The grid is rectangular, with x axis generally perpendicular to shore and y axis parallel to shore. Subscripts i and j index rows and columns of the grid and corresponding arrays, and also serve as x and y coordinates. The grid is augmented by two additional columns $N+1$, $N+2$ (Figure 3-12) that facilitate a "leapfrog" finite-difference procedure where values at each grid point are partly determined from values at surrounding grid points. Additional columns provide cells required for finite-difference calculations at grid cells that are based on values in four neighboring cells. Conventions

66

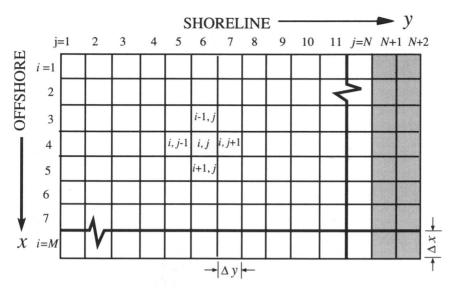

Figure 3-12 Grid used in WAVE to represent coastline. Rows and columns are indexed with *i* and *j* that also serve as *x* and *y* coordinates, respectively. Cells immediately adjacent to cell *i, j* (dark shaded) are indexed by incrementing or de-incrementing *i* and *j* by 1. Augmented columns *N* and *N+1* (shaded) are required for finite-difference solution scheme. *M* is maximum number of rows and *N* is maximum number of columns in grid representing geographic area of simulation.

for describing wave angles and longshore currents within WAVE's grid scheme are shown in Figure 3-13.

WAVE's finite-difference scheme

Once an augmented grid scheme is established, governing wave equations including those for satisfying continuity (Equation 3-55), momentum (Equations 3-64, 3-65), and wave number (Equation 3-44) are transformed into finite-difference approximations provided by Ebersole and Dalrymple (1979). Finite-difference approximations of continuity and momentum equations can be expressed using an abbreviated central-difference scheme for time-dependent terms (Ebersole and Dalrymple, 1979, p. 44-47), where numerical solutions of wave parameters during subsequent iterations *n+1* are calculated from values obtained in the two immediately previous iterations *n*, and *n* 1, where the time difference between iterations *n*, *n* 1, and *n+1* is two time steps, or *2Δt*:

$$\eta_{i,j}^{n+1} = \eta_{i,j}^{n-1} + 2\Delta t F_1^{n,n-1} \qquad (3\text{-}75)$$

$$U_{i,j}^{n+1} = A U_{i,j}^{n-1} + 2\Delta t F_2^{n,n-1}$$

$$V_{i,j}^{n+1} = B V_{i,j}^{n-1} + 2\Delta t F_3^{n,n-1}$$

67

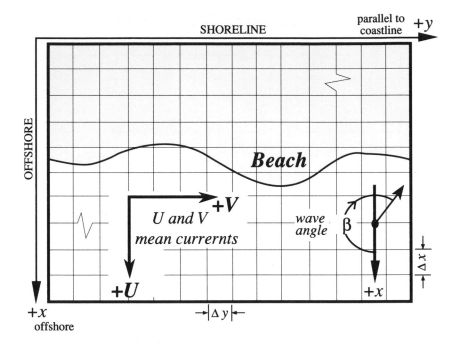

Figure 3-13 Maps showing scheme for representing area of hypothetical beach and algebraic sign conventions for currents and wave angles used by WAVE. U and V represent components of longshore currents and angle β represents angle of wave approach.

where

η = instantaneous surface elevation of wave, or setup

n = time step number

U = velocity in x direction (Figure 3-13)

V = velocity in y direction (Figure 3-13)

Δt = time step

A, B = functions of depth, where values were evaluated at previous time step, at different depth $d+\eta$

i, j = subscripts corresponding to x and y axes (Figure 3-12)

F_1, F_2, F_3 = various functions or variables within Equations 3-55, 3-64, and 3-65, respectively

Functions A and B are solely functions of water depth, whereas F_1, F_2, and F_3 are functions of many variables that control the shoaling character of waves. During the first iteration, values for functions A, B, F_1, F_2, and F_3 are solved using a forward-difference procedure (Figure 3-14A), where a "first-guess" of values of η, U and V are calculated for the present time step n, using values of A, B, F_1, F_2,

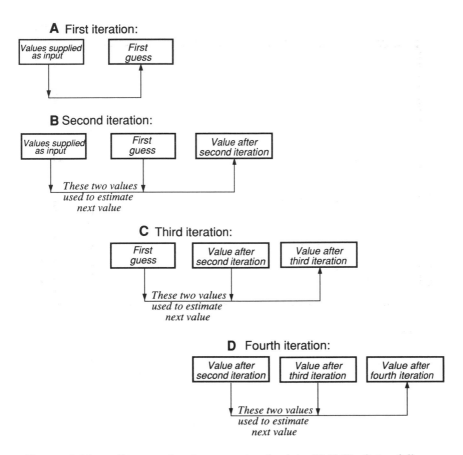

A First iteration:

| Values supplied as input | First guess |

B Second iteration:

| Values supplied as input | First guess | Value after second iteration |

These two values used to estimate next value

C Third iteration:

| First guess | Value after second iteration | Value after third iteration |

These two values used to estimate next value

D Fourth iteration:

| Value after second iteration | Value after third iteration | Value after fourth iteration |

These two values used to estimate next value

Figure 3-14   Diagram showing steps involved in WAVE's finite-difference "leapfrog" scheme for obtaining wave parameters: (A) During first iteration, WAVE uses forward difference where values supplied as input are used to estimate first guess of wave parameters. (B) Values of next iteration are calculated with leapfrog scheme, where values are calculated from those of two previous time steps. (C) Third, (D) fourth, and subsequent iterations continue to use leapfrog scheme, where values at each subsequent time step are calculated from two immediately preceding time steps.

and F_3 provided as input for the initial conditions. New values for η, U and V are then used to solve wave equations for changes in wave angles, wave heights, and wave numbers as waves move toward shore. The process is repeated with the forward finite-difference scheme providing solutions during each new time step n from values of previous time steps. Wave characteristics are calculated only for submerged cells, and cells whose elevations are above sea level are ignored.

Subsequent iterations use a "leapfrog" scheme (Figure 3-14 B, C and D), where η, U, and V calculated during current time step n and previous time step $n-1$, yield values for the next time step $n+1$. The name "leapfrog" pertains to use of

69

information from the two preceding time steps. The first forward-difference iteration provides a rough estimate of wave parameters throughout the grid, yielding values that the leapfrog solution scheme then utilizes in the next iteration. Figure 3-14 shows the leapfrog procedure during subsequent iterations, where values at time $n+1$ are based on values at times n and $n-1$.

Program execution

Figure 3-15 summarizes the organization of WAVE's circulation module WAVECIRC, with roles of WAVECIRC's subroutines listed in Table 3-1. WAVE begins by calculating the characteristics of a shoaling wave front advancing through the grid. Initially water is at rest and velocity components throughout the grid are zero. Snell's law provides the initial estimate of wave heights and refraction angles during the first iteration over grid (Figure 3-16). Subsequent iterations use finite-difference schemes to update values, providing new wave angles θ, wave numbers k, wave heights H, surface elevations η, and current velocities U and V. Additional iterations increase the accuracy of solutions. Wave heights are gradually increased from zero to their full deep-water value over a specified number of iterations to provide a gradual increase in wave heights and increased numerical stability. Gradual increase prevents "shock loading," and is analogous to gradual generation of waves in a wave tank instead of sudden propagation of a large wave front.

Table 3-1   Roles of subroutines in WAVECIRC.

Subroutine	Role
DGRAD	Determines local water depth $\eta+D$
REFRACT	Calculates refraction and radiation stresses
SNELL	Calculates first-guess of wave angle θ
WVNUM	Calculates wave number k
ANGLE	Calculates wave angle θ
HEIGHT	Calculates wave height H
GROUP	Calculates wave celerity C and group velocity C_g
TAUSB	Calculates surface and bottom shear stresses τ_s and τ_b
CONTIN	Solves continuity equation for η
ETAS	Calculates change in water depth $\eta+D$
MOMEN	Solves momentum equations for U and V
UCALC	Calculates depth-averaged values for U and V
ROLBAC	Updates η, U, V, and D for next iteration

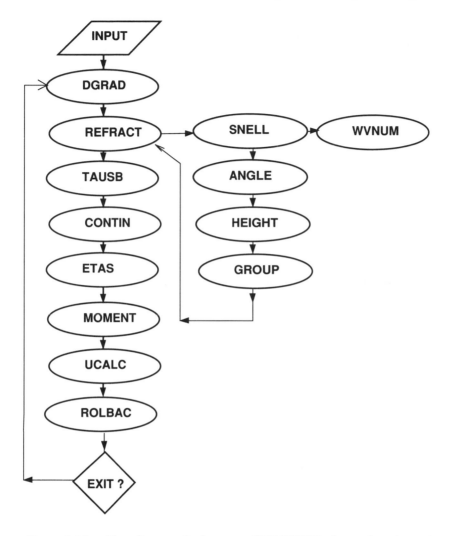

Figure 3-15   Flow diagram of subprogram WAVECIRC whose subroutines are described in Table 3-1. Arrows indicate sequence in which subroutines are executed. WAVECIRC's function within WAVE is shown in Figure 5-7.

Sensitivity of wave parameters to grid size and input parameters

Wave parameters provided by WAVECIRC are sensitive to grid size, time steps, and other data provide as input. Experiments involving WAVECIRC are monitored by plotting wave-induced velocities U and V calculated during an experiment. For example, onshore-offshore and longshore currents can be plotted versus number iterations (Figure 3-17) to show when currents converge to their steady state values. However, grid cell size affects the time step in WAVECIRC's

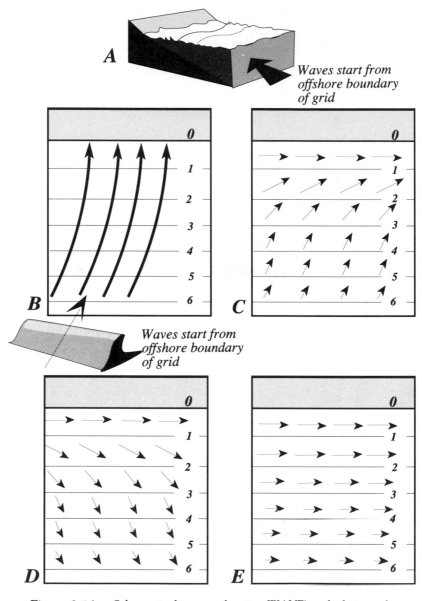

Figure 3-16   Schematic diagrams showing WAVE's calculation of wave-induced currents: (A) Perspective diagram of planar beach (B) Contour map of depths (in meters) showing that wave crests propagated over area cause wave orthogonals (curved bold arrows) to refract so that they tend to arrive perpendicular to shore. (C) Map of onshore component of currents represented by vectors plotted cell-by-cell over grid. (D) Offshore components. (E) Net depth-averaged, time-averaged longshore currents representing average of onshore and offshore currents. Contours show depths in meters.

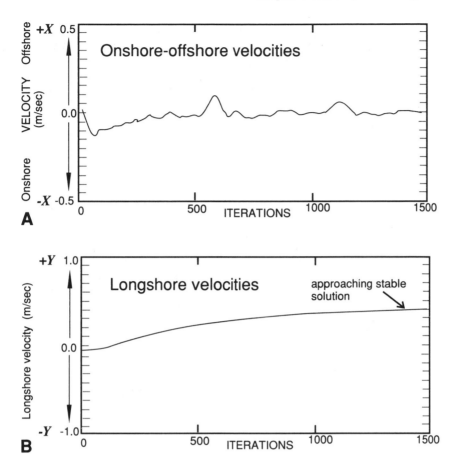

Figure 3-17 Plots of velocities versus number of iterations to determine when stable numerical solutions have been obtained. Experiment involves both onshore-offshore currents and longshore currents computed by WAVE after 1500 iterations. Plots pertain to currents in cell {3,3} in Figure 3-20: (A) Onshore-offshore velocities versus iterations. (B) Longshore velocities versus iterations, with convergence after about 1500 iterations.

numerical solutions, and in turn affects the number of iterations required to obtain steady state solutions for wave parameters.

Figure 3-18 shows example results of three experiments involving WAVECIRC, where wave-induced currents were calculated for an area spanning 1000 by 1000 meters, having a beach slope of 0.025. In the first experiment the simulated area was divided into 100 cells (10 rows and 10 columns), each 100 meters square. In the second experiment, 400 cells (20 rows and 20 columns), each 50 meters square, were used to represent the same area. In the the third experiment, 1000 cells (40 rows and 40 columns), each 25 meters square, were used. In this way, an area

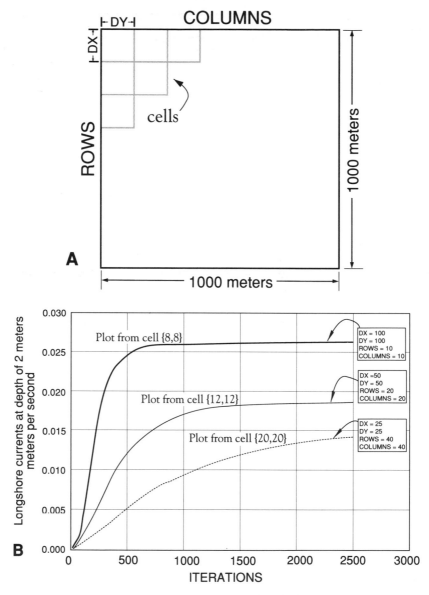

Figure 3-18 Map of grid area, and plots of longshore currents showing sensitivity of currents to grid spacings during three experiments involving same geographic areas, but employing different numbers of grid cells: (A) Map of grid area 1000 meters square, showing division of cells into rows and columns. Experiments employ 10 rows and 10 columns, 20 rows and 20 columns, and 40 rows and 40 columns, respectively, to represent area. (B) Plot of longshore currents versus number of iterations for three experiments involving varying numbers of cells used to represent similar geographic areas. Plots are from cells having similar water depths. Steady state values for currents are obtained when plots become nearly horizontal.

spanning 1000 by 1000 meters was represented by three grids of varying discretization.

Longshore currents in Figure 3-18B are plotted from cell 8,8 in the first experiment, cell 12,12 in the second experiment, and cell 20,20 in the third experiment, each of which had similar water depths. Figure 3-18 shows that currents converge quickest and are greatest in the 10-by-10 grid, and longshore currents are similar for all three, suggesting that WAVECIRC's numerical procedures can be employed to estimate currents in grid schemes with cells of various sizes. Similar sensitivity tests performed for other geographic areas (Figure 3-19) show that WAVECIRC converges fastest when fewer cells are used, and that more iterations are required if there are more cells in the grid. Other experimental results, including directions of longshore currents and wave angles, were monitored by plotting them as vectors on contour maps that represent topography (Figure 3-20). Martinez (1987b) provides additional discussion of WAVECIRC's performance at varying geographic scales and of the sensitivity of simulated currents to various input parameters.

Other computer procedures could be substituted within WAVE, but they must predict refraction angles, breaking-wave heights, wave-induced currents, orbital velocities, wave-induced shear stresses, radiation stresses, and other parameters required for simulating sand transport by waves (Figure 3-21). Improved procedures for shoaling and refraction of waves in three dimensions could better represent currents and shear stresses produced by waves that transport sediment. Until then, procedures for simulating sediment transport will be limited by simplifications incorporated by WAVECIRC, where near-bottom currents and shear stresses are approximated by depth-averaged values. Despite these simplifications, hydrodynamic parameters provided by WAVECIRC are generally adequate for use with equations in Chapter 4 for calculating sediment transport rates.

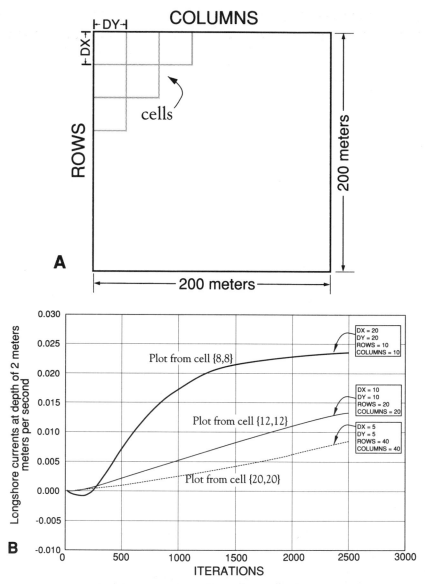

Figure 3-19 Map of grid area and plots of longshore currents showing sensitivity of currents to grid spacings during three experiments involving same geographic areas but employing different numbers of grid cells: (A) Map of grid 200 meters square, showing division of cells into rows and columns. Experiments employ 10 rows and 10 columns, 20 rows and 20 columns, and 40 rows and 40 columns of cells, respectively, to represent simulation area. (B) Plot of longshore currents versus number of iterations plotted for three experiments. Plots are from cells having similar water depths. Steady-state values for currents are obtained when plots become nearly horizontal.

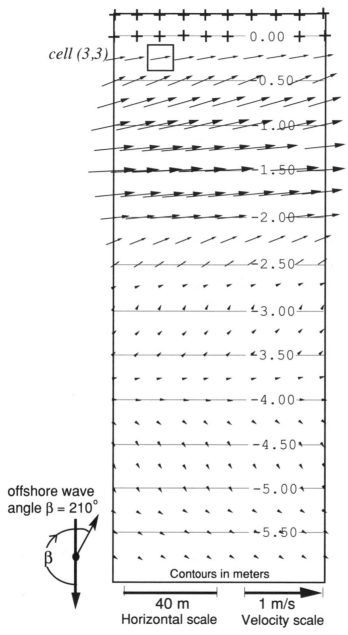

Figure 3-20 Contour map of water depth in meters showing wave-induced currents computed by WAVE during experiment involving 1500 iterations shown in Figure 3-17. Vectors represent net time-averaged, depth-averaged wave-induced currents. Crosses represent areas above sea level. Plots in Figure 3-17 were produced from currents calculated for cell{3,3} shown by box.

77

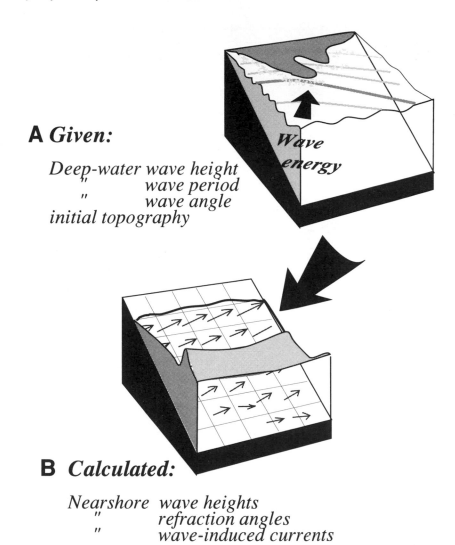

A *Given:*

Deep-water wave height
"   wave period
"   wave angle
initial topography

B *Calculated:*

Nearshore wave heights
"   refraction angles
"   wave-induced currents

Figure 3-21   Diagram comparing information provided as input to WAVECIRC to parameters calculated by WAVECIRC: (A) Initial topography and deep-water wave height, wave period, and deep water wave angle are provided as input. (B) WAVECIRC computes shallow-water wave heights, refraction angles, and wave-induced currents required to represent sand transport by waves. Values are assigned to grid cells.

CHAPTER 4

Nearshore Sediment Transport

Many equations have been proposed for calculating nearshore sediment transport, but some are too complex for incorporation in computer procedures and most are limited in their application to specific parts of the nearshore environment. Thus, selection of appropriate equations is difficult and only a few of the many alternative equations are described below. WAVE's equations for representing sediment transport have been selected to accord with field observations that show a strong correlation between longshore transport rates and wave energy. This chapter reviews early field studies that established the correlation between longshore transport and wave energy, introduces mechanisms of sand transport in offshore and nearshore areas, and introduces general types of equations for representing mechanisms of transport which vary between nearshore and offshore areas. These transport equations are described, along with assumptions involved in their formulation and the rationale for their selection.

The most widely used equations for estimating transport rates are based on studies by Watts (1953), Caldwell (1956), and Komar (1969), that are summarized in Tables 4-1 to 4-5, and Tables A-1 to A-7 in Appendix A). Observations in these studies show a strong correlation between longshore transport rate and wave power. Greer and Madsen (1978) and Komar (1988, 1991) reviewed these studies and discussed their assumptions and shortcomings. These studies are important not only because they provide the basis for WAVE's transport equations, but also because they provide data for comparison with simulation experiments presented in Chapters 6 and 7.

Watts (1953) was one of the first to document the relationship between wave power and longshore transport rates. His data were used by the U. S. Corps of Engineers to establish one of the first equations for estimating rates of longshore transport. Watts

Table 4-1 Field studies of beaches where wave-power equation was used to estimate coefficient K (used in Equation 4-17) to estimate sand transport rates. Modified from Komar (1988).

Reference	Location	Mean grain sz in mm	Coefficient K: Mean value and observed ranges
Watts (1953)	S. Lake Worth Inlet, Fl.	0.40	0.89 (.73-1.03)
Caldwell (1956)	Anaheim, California	0.40	0.63 (.16-1.65)
Komar and Inman (1970)	El Moreno, Mexico	0.60	0.82 (.48-1.15)
Komar and Inman (1970)	Silver Strand, Calif.	0.18	0.77 (.52-.92)
Lee (1975)	Lake Michigan	?	0.42 (.24-.72)
Knoth and Nummedal (1977)	Bull Island, S. Carolina	0.18	0.62 (.23-1.0)
Inman, and others (1980)	Torrey Pines, Calif.	0.20	0.69 (.26-1.34)
Duane & James (1980)	Pt. Mugu, Calif.	0.15	0.81
Bruno and others (1981)	Channel Is. Calif.	0.20	0.87 (0.42-1.5)
Dean and others (1982)	Santa Barbara, Calif.	0.22	1.15 (.32-1.63)
Dean and others (1987)	Rudee Inlet, Virginia	0.30	1.00 (.84-1.09)

Table 4-2 Average wave and beach conditions at South Lake Worth inlet. After Watts (1953).

Length of beach studied	500 m
Grain diameter	0.39 mm north of jetty, at mean tide line
	0.58 mm south of jetty, at mean tide line
Grain density	2650 kg/m^3
Breaker height	0.25 m at depth of 5.2 m
Wave period	5.0 sec
Longshore currents	0.20 m/s
Alongshore energy	93 watts/m
Volume transport	260 m^3 /day

Table 4-3 Average wave and beach conditions at Anaheim Beach. After Caldwell (1956).

Length of beach studied	5000 m
Beach slope	0.007 steepening toward shore
Grain diameter	0.40 mm, varying with distance from jetty
Grain density	2650 kg/m^3
Breaker height	0.5 m where depth is 3.5 m
Wave period	16 sec
Incident angle	5 degrees, varying along beach
	northern swell : 10 degrees; southern; 20
Width of surf zone	75 m, estimated from aerial photos
Alongshore energy	440 watts/m
Volume transport	854 m^3 /day

Table 4-4 Average wave and beach conditions at El Moreno Beach. From Komar (1969) and Komar and Inman (1970).

Length of beach studied	125 m
Beach slope	0.135
Grain diameter	0.60 mm within surf zone
Grain density	2650 kg/m^3
Breaker height	0.3 m
Wave period	3 sec
Breaker type	spilling
Incident angle	9 degrees
Width of surf zone	17 m
Sand advection	0.0019 m/sec
Depth mobile bed	0.084
Longshore currents	0.3 m/s
Orbital velocities	0.8 m/s at breaking
Rip spacing	15 m
Tidal influence	neglible compared to longshore currents
Wind	generally mild onshore sea breeze
Alongshore energy	30 watts/m
Volume transport	225 m^3 /day

Table 4-5 Average wave and beach conditions at Silver Strand Beach. From Komar (1969) and Komar and Inman (1970).

Length of beach studied	125 m
Beach slope	0.034
Grain diameter	0.175 mm in surf zone
Grain density	2650 kg/m^3
Breaker height	0.7 m
Wave period	11 sec
Breaker type	plunging
Incident angle	5 degrees
Width of surf zone	75 m
Sand advection	0.0023 m/sec
Depth mobile bed	0.04 m
Longshore currents	0.2 m/s
Orbital velocities	1.2 m/s at breaking
Rip spacing	300 m
Wind	waves generated from swell or storms
Alongshore energy	50 watts/m **
Volume transport	280 m^3 /day **

**Excludes one extreme value (see Table A-7 in Appendix A)

studied transport rates at South Lake Worth inlet, Florida by measuring volumes of sand pumped past a jetty by a nearby pumping station (Figure 4-1). When in operation, the pumping station kept sand from accumulating against the jetty and Watts assumed that the volume of pumped sand roughly equaled the longshore transport rate. Watts measured deep-water wave heights and wave angles, and used Airy wave theory to calculate incoming wave power. With ideas proposed by Munch-Petersen (1950), Watts plotted transport rate Q with longshore wave power P_l, yielding a graph similar to Figure 4-2.

Caldwell's (1956) study provided additional data that helped confirm the correlation between sand transport and wave energy. Caldwell measured the movement of several thousand cubic meters of sand placed on the downdrift side of the jetties at Anaheim Bay, California (Figures 1-2 and 4-3). Assuming that the sand was completely blocked updrift of the jetties, Caldwell estimated sediment transport rates by measuring changes in submerged topography downdrift of the jetties, employing 21 transects oriented perpendicular to the beach and spaced approximately 153 meters apart (Figure 4-3B). Wave power and other parameters in the vicinity (Table 4-3) were determined from wave gauges and hindcast data. Similar to Watts, Caldwell plotted transport rate Q versus longshore wave power P_l, yielding a graph similar to Figure 4-2, and documenting the correlation between transport rate and wave power.

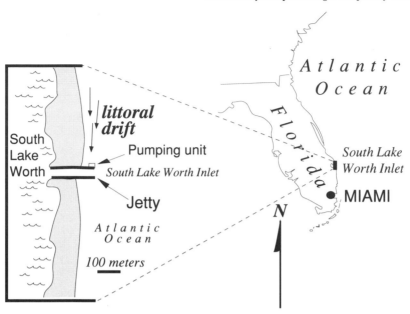

Figure 4-1 Map of South Lake Worth inlet and jetty. Modified from Watts (1953).

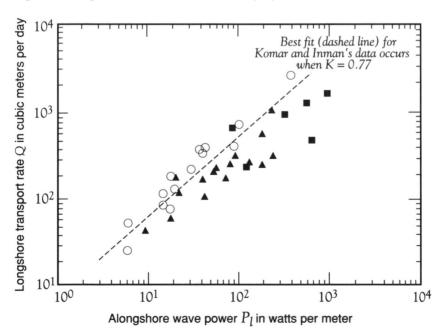

Figure 4-2 Log-log plot of sediment transport rate Q versus wave power P_l based on field studies by Watts (1953) [triangles], Caldwell (1956) [squares], and Komar and Inman (1970) [circles]. Line represents best fit to Komar and Inman's data, where coefficient K in wave-power equation (Equation 4-17) is 0.77. Data for plot from Tables A-2, A-4, and A-7 of Appendix A.

83

Komar and Inman (1970) estimated longshore transport rates at El Moreno Beach in Baja California, Mexico and Silver Strand Beach in southern California (Figure 4-3A) by measuring movement of tracers consisting of sand grains stained with

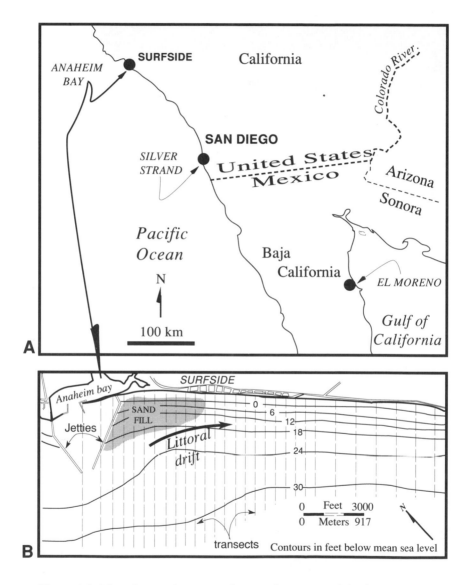

Figure 4-3 Maps showing location and geographic setting of Anaheim Bay jetty: (A) Location map showing Anaheim Bay, El Moreno, and Silver Strand beaches (B) Contour map of submerged topography near Anaheim Bay jetties. Contours show depth in feet with respect to sea level. Shading shows location of sandfill downdrift of jetty. Dashed lines show location of transects where topography was monitored. After Caldwell (1956).

fluorescent dyes (Figure 4-4). Similar to Watts and Caldwell, they also found that transport rates are correlated with wave power (Figure 4-2). Additionally, they proposed a theoretical basis for the correlation and devised new equations for estimating transport rates (Equations 4-14 to 4-17). These are the equations incorporated in WAVE.

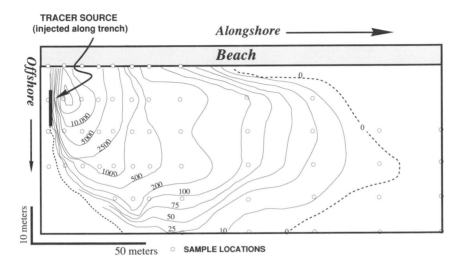

Figure 4-4 Map showing concentration of tracers in sand at El Moreno Beach. Contours represent concentration of tracers in grains per kilogram of sand. Dyed sand grains that served as tracers were supplied in trench and concentrations were measured four hours after their release. After Komar (1969).

PROCEDURES FOR REPRESENTING TRANSPORT BY WAVES

Nearshore transport includes sediment moved both by longshore currents generally parallel to shore, and onshore-offshore transport moving perpendicular or oblique to shore. Accordingly, mathematical or computer procedures devised to represent transport are separated into two general categories, namely, those that represent longshore transport and those that represent onshore-offshore transport (Figure 2-2). Procedures for representing onshore-offshore transport generally apply seaward of the breaker zone, where onshore and offshore wave-motions are generally parallel to the direction of wave propagation (Figures 2-2 and 2-7). By contrast, procedures for representing longshore transport generally apply shoreward of the breaker zone, where currents are generally steady and move sand parallel to shore (Figures 2-2 and 2-8). Few equations or groups of equations effectively represent transport in both breaker and swash zones, and few if any adequately represent sediment transport in both offshore and nearshore areas. Combining equations for oscillatory, turbulent, and steady

85

currents into a group of computer procedures is not yet feasible and will require full three-dimensional representation of fluid motions. As a result, a specific computer program may generally be suited for simulating either offshore areas, or nearshore areas, but rarely can a single program represent all of the processes affecting both (Figure 4-5).

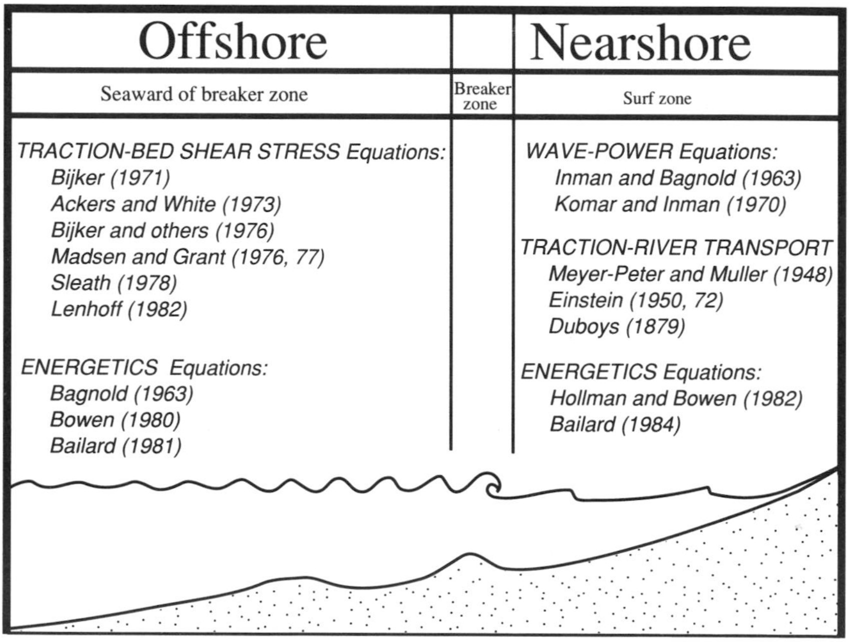

Offshore		Nearshore
Seaward of breaker zone	Breaker zone	Surf zone

TRACTION-BED SHEAR STRESS Equations:
  Bijker (1971)
  Ackers and White (1973)
  Bijker and others (1976)
  Madsen and Grant (1976, 77)
  Sleath (1978)
  Lenhoff (1982)
 	
ENERGETICS Equations:
  Bagnold (1963)
  Bowen (1980)
  Bailard (1981)
		WAVE-POWER Equations:
		  Inman and Bagnold (1963)
		  Komar and Inman (1970)
		
		TRACTION-RIVER TRANSPORT
		  Meyer-Peter and Muller (1948)
		  Einstein (1950, 72)
		  Duboys (1879)
		
		ENERGETICS Equations:
		  Hollman and Bowen (1982)
		  Bailard (1984)

Figure 4-5 Schematic cross section showing where types of transport-equations are applied in nearshore versus offshore areas.

Procedures for representing onshore-offshore transport

   In deep water, orbital wave motions do not affect the sea bottom, but in shallow water waves do have an effect as they begin to "feel" the bottom and their orbital motions lift, suspend, transport, and sort sediment (Figures 3-4 and 3-6). For example, larger grains tend to move as bedload near the bottom (Figure 1-4), and they may be moved shoreward by strong velocities, but return velocities may be too low to move them back again (Figure 4-6). Lesser particles tend to move higher in the water column as suspended load, and may be moved back and forth, with slow movement in the direction of wave propagation. The finest particles, however, tend to move back and forth in suspension within wave orbitals and can move offshore by gravity. No distinct boundary exists between bedload and suspended load, and for convenience a boundary is generally assumed to be a few centimeters above the bottom. McCave (1972), Coakley and Skafel (1982), Sternberg and others (1984,1989), and Zampol and Inman

(1989), describe procedures for measuring suspended load and Ingle (1966), and Komar (1969), Sleath (1978) and Niederoda and others (1982) describe procedures for measuring bedload transport.

These generalizations about onshore and offshore sediment transport due to oscillatory wave motion are simplified. The net cross-shore transport of sediment by oscillatory wave motions is seldom observed, and if it actually occurred, it would tend to produce either continuously accreting or eroding beaches. Instead, onshore and offshore transport generally reach an equilibrium so that an *equilibrium beach profile* is formed where the onshore transport of larger grains is balanced by the offshore transport of smaller grains.

The development of an equilibrium beach profile can be explained through the null-point concept developed by Cornaglia (1950), Ippen and Eagleson (1955), Miller and Zeigler (1958), and Eagleson and Dean (1961). According to the null-point theory, grains of different size have unique equilibrium position along a beach slope, where they either remain unmoved or are suspended without net movement either shoreward or seaward. Grains tend to migrate toward those equilibrium positions. If there is enough time an entire beach may achieve equilibrium, with little change in submerged topography even though transport of sediment continues. Statistical correlation between grain size and beach slope (Bascom, 1951, 1954, 1959) supports the null-point theory, but beaches are rarely if ever in equilibrium. Instead, they adjust to hourly, daily, and seasonal changes in wave climate and sediment supply (Figures 2-4, 2-5, and 2-6).

Onshore-offshore transport is difficult to represent with computer procedures because currents are non-steady over wave periods spanning a few seconds, and rates and directions of sediment transport fluctuate during the passing of individual waves (Figure 4-6). Some procedures or equations are described by Bagnold (1963, 1966), Bijker (1971), Einstein (1972), Ackers and White (1973), Swart (1974), Bijker and others (1976), Madsen and Grant (1976, 1977), Sleath (1978), Van de Graaf and Van Overeem (1979), Bowen (1980), Felder and Fisher (1980), Van de Graaff and Tilmans (1980), Bailard (1981), Lenhoff (1982), Bowen and Doering (1984), Richmond and Sallenger (1984), Noda (1984), Stive and Battjes (1984), Leontev (1985), Martinez (1987a,b), Lee-Young and Sleath (1988), Seymour and Castel (1989), and Swain (1989). However, these procedures are generally two dimensional, with experimental results generally similar to those shown in Figure 2-6, where evolving beach profiles are shown by multiple cross sections. King and Seymour (1989) provide a concise review of the most widely used procedures, while Horikawa (1988) tabulates and compares equations used by various procedures. While satisfactory for representing onshore-offshore transport caused by orbital motions of waves, these procedures are severely limited by their two-dimensional application and the short spans of time that are simulated.

Representing longshore transport

Longshore currents are dominant shoreward of the breaker zone and generally move sediment parallel to shore, while rip currents move sediment offshore

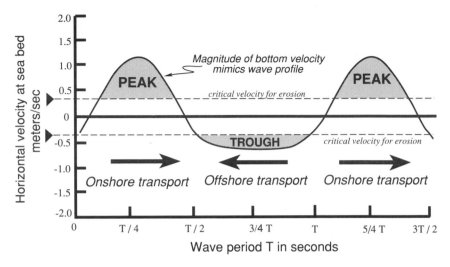

Figure 4-6 Diagram contrasting shoreward transport with offshore transport of sediment. Volume of sediment transported in onshore direction may differ from offshore direction during passage of wave. Bottom velocities under wave peaks are stronger, but of shorter duration, than under wave troughs.

(Figures 2-2, 2-8, and 2-9). Although longshore and rip currents are strongest in the surf zone, other currents also transport sediment in the nearshore area. For example, bores (Figures 2-2 and 2-3) created by reformed waves cause hydraulic jumps that momentarily suspend and transport sediment shoreward, while reflected waves from the swash zone move sediment offshore. Ingle (1966) observes that sediment generally moves alongshore within the breaker zone, while in the swash zone, Komar (1969, 1971) observes that sand moves in a saw-tooth pattern. Undertow or return flows (Figure 2-8B) associated with breaking waves are also important for offshore sediment transport from the nearshore area (Stive and Battjes, 1984; Miller and Zeigler, 1964).

Procedures for representing longshore transport are usually quasi three-dimensional where evolving topographic features or the position of the shoreline are merely displayed with two dimensional contour maps of the surface topography (Figure 4-7), although additional procedures can be included to display the three dimensional record of sedimentary deposits (Plates 4 and 5). Longshore transport is readily represented with computer procedures because currents and sediment transport are assumed to be generally steady within the surf zone and longer spans of time can be simulated without recalculating current fields. Some procedures for representing longshore transport have been described by Einstein (1948, 50, 72), Munch-Petersen (1950), Inman and Bagnold (1963), Komar (1969, 1971, 73, 76, 88, 91), Komar and Inman (1970), Fleming and Hunt (1970), Swart (1976), Ozasa and Brampton (1980), Holman and Bowen (1982), Bailard (1984), Kamphuis and Sayao (1982), Kamphuis and others (1986), DeVriend (1987), Kraus and others (1988). While these procedures are generally satisfactory for representing longshore transport caused by steady longshore

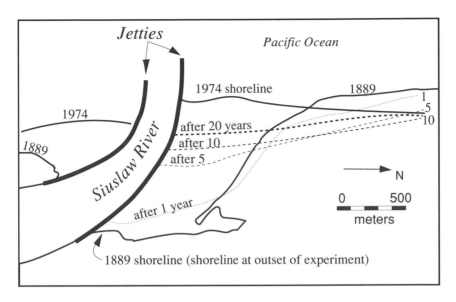

Figure 4-7 Map of simulated sand transport around jetties at Siuslaw River, Oregon (Komar, 1977). Positions of actual shoreline in 1889 and 1974 are shown for comparison. At outset of simulation experiment, shoreline was placed where actual shoreline was located in 1889. Shifts of simulated shoreline are shown after 1, 5, 10, and 20 years.

currents, they are generally unable to represent fluctuating transport rates caused by orbital motions of waves, nor do they record a three-dimensional representation of sediment moved by waves. Unlike most computer models, WAVE does maintain a three-dimensional record of sediment deposited by waves, but its use of time-averaged and depth-averaged currents still limits its ability to represent transport by nonsteady currents.

Procedures for representing both onshore-offshore and longshore transport
Procedures have been proposed to represent both onshore-offshore and longshore transport, thereby representing transport by both steady and unsteady currents. These procedures generally incorporate an extra coefficient for friction (Jonsson, 1966) that empirically represents the additional shear stresses imparted by oscillatory wave motions. Bailard (1984) and Holman and Bowen (1982) provide examples of procedures that incorporate aspects of oscillatory and longshore flow, and Felder and Fisher (1980) developed a unique, hybrid procedure that employs different sediment transport equations in offshore and nearshore areas. Three-dimensional representation of flow is required to represent combined oscillatory and longshore currents, so their representation with computer procedures is complicated (Miller and Zeigler, 1964; Bijker and others, 1976; Bakker and Doorn, 1978; Stive and Battjes, 1984; Dean and Perlin, 1986; Kim and others, 1986; Davies, 1987; Ramsden and Nath, 1988; and

89

Whitford and Thornton, 1988). Holman (1991, personal communication) suggests that onshore-offshore transport is still too poorly understood to be adequately simulated in three dimensions and recommends that computer procedures focus on representation of longshore transport. Thus, while procedures for representing the combined effects of longshore, onshore, and offshore transport have not achieved wide application, the next generation of procedures for simulating nearshore transport probably will represent mechanisms of sand transport that vary between offshore and nearshore areas.

EQUATIONS FOR NEARSHORE TRANSPORT

Equations for representing nearshore transport are largely empirical, which is not surprising considering the complexity of nearshore environments which are exposed to oscillatory, refracting, and reflecting waves of variable intensity. Nearshore areas are also affected by variable shelf currents and tidal forces and must absorb or transmit energy supplied by incoming waves. Nearshore systems are therefore more difficult to represent than stream systems, which can be approximated with the Navier-Stokes equations and other related equations that represent flow in open channels. It is also more difficult to gather data along coastlines and most theoretical work is based on the few field studies listed in Table 4-1. Because of the relative paucity of data for coastlines, equations for nearshore sediment transport have been adapted largely from studies of sediment transport by streams (DuBoys, 1879; Shields, 1936; Bagnold, 1940; 1956, 1962, 1963, 1966; Einstein 1948, 1950; and Meyer-Peter and Muller 1948), except that additional terms are employed to represent the effects of waves. Equations for transport rate Q_s in the nearshore area are generally reported in cubic meters per day or cubic meters per year, and generally include many interdependent parameters:

$$Q_s = f(H, L, T, d, \rho_s, \rho_w, \mu, g, x, y, z, t, D, \tau_s, \tau_b, \alpha, \tan\beta) \qquad (4-1)$$

where

H	=	wave height	μ	= fluid viscosity
L	=	wavelength	g	= gravitational acceleration
T	=	wave period	x, y, z	= spatial coordinates
d	=	water depth	t	= time
ρ_s	=	density of sediment	D	= average grain diameter
ρ_w	=	density of water	τ_s	= surface shear stress
τ_b	=	bottom shear stress	$\tan\beta$	= beach slope
α	=	angle of wave incidence		

While all equations for representing transport by waves are empirical, some are more empirical than others. In general, equations derived from physical principles are much less empirical than those from statistical analysis, although selection of appropriate equations is difficult because the physics of sediment transport by waves cannot be adequately described by a few simple equations. In fact, hundreds of equations have been proposed for describing sediment transport by waves (Horikawa, 1988). Many share basic physical principles or assumptions and are grouped here into four classes designated here as "wave-power", "traction", "energetics", and "empirical" equations.

Wave-power equation

The wave-power equation assumes that sediment transport rates are proportional to the power or energy per unit of time, of waves impinging on the coast (Figure 4-8). The wave-power equation was first proposed by the U. S. Army Corps of Engineers (Munch-Petersen, 1950) and is called the "SPM equation" after the *Shore Protection Manual* (U.S. Army CERC 1977) when expressed in the following form:

$$Q = K_1 E_o \cos \alpha \qquad (4\text{-}2)$$

where

Q = volumetric rate of sediment transport (cubic meters per second)
K_1 = calibration coefficient
E_o = wave energy density per unit width of wave crest, based on Equation 3-63, evaluated in deep water
α = angle of incidence of approaching waves (denoted as β in Equation 3-31)

Field data of Watts (1953) and Caldwell (1956) and others listed in Table 4-1 permit calibration of K_1 in Equation 4-2, whose simplicity and success in prediction has made it widely used for determining longshore transport rates. Although empirical and derived from both intuition and observation, Inman and Bagnold (1963) demonstrate its soundness by presenting a similar derivation where sediment transport rate is related not only to wave power but also to the ratio between average longshore velocity v_l and maximum orbital velocity u_{max} :

$$Q = K_2 (E_b C_b) \cos \alpha_b (v_l / u_{max}) \qquad (4\text{-}3)$$

where

K_2 = calibration coefficient
E_b = wave energy density per unit width of wave crest, at breaker zone (Equation 3-63)
C_b = group velocity at breaker zone (Equation 3-68)
v_l = average longshore velocity (Equation 3-65)
u_{max} = maximum orbital velocity near sea bottom (Equation 3-26)
α_b = angle of incidence at breaker zone

91

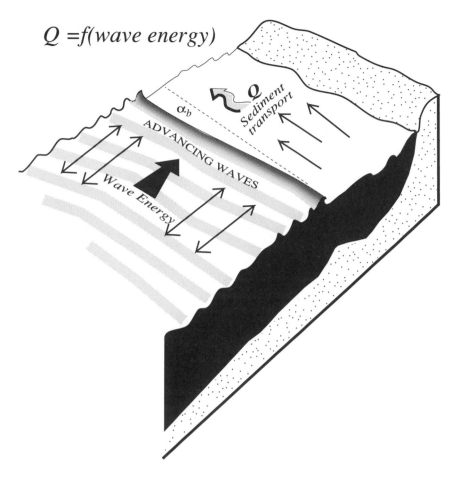

Q =f(wave energy)

Figure 4-8   Advancing waves perform work and drive nearshore circulation system that transports sand alongshore. Wave-power equation of Inman and Bagnold (1963) and Komar and Inman (1970) relates sediment transport rate Q to amount of energy provided by incoming waves. Longshore currents and related littoral transport are also functions of incident angle α_b. Resulting radiation stresses are dissipated alongshore.

Komar and Inman (1970) propose a similar derivation (Equation 4-4), where longshore transport rate is a function of wave energy and the angle of incidence of breaking waves α_b:

$$Q = K_3 \ (E_b C_b) \cos \alpha_b \sin \alpha_b \qquad (4\text{-}4)$$

where

K_3 = calibration coefficients

Their equation treats longshore transport rates as a function of power available at the moment that waves break. While it can be used to determine longshore transport rates within the surf zone, it is not useful for representing onshore-offshore transport offshore from the breaker zone. The equation, which is included in WAVE, is widely used and is simple to solve analytically. Calibration of coefficient K_3 is based on numerous field studies, and other parameters in Equation 4-4 are easily measured in field studies. Vincent (1979), Carmel and others (1984), and Chandramohan and others (1988) provide examples where wave-power equations have been applied at large geographic scales.

Traction equations for sediment transport

Traction equations incorporate basic physical principles and include familiar parameters such as Shields parameter Ψ, which although empirical, yields the quantitative threshold velocity for grain movement. Traction formulations include bed shear stresses, and gravitational, hydraulic, hydrostatic, and buoyant forces that act on individual grains or on masses of moving bedload (Figure 1-4). Traction equations generally assume that flows are steady, a condition generally not met in nearshore areas, where oscillatory motions of waves create near-bottom velocities that change direction, accelerate, and deaccelerate with each passing wave. Traction equations have been applied where oscillatory currents are averaged during a single wave period, but these equations are difficult to encode as computer programs and require calibration of coefficients that are poorly understood and not directly available from field studies. Unlike the wave-power equation, no single traction equation has gained widespread use, and the difficulty in estimating onshore and offshore transport seaward of the breaker remains.

Traction equations express sediment transport rate Q as a function of bed shear stress or Shields parameter Ψ (Madsen and Grant, 1976):

$$Q_{(t)} = k\Psi_{(t)} \tag{4-5}$$

where

$Q_{(t)}$ = instantaneous volumetric sediment transport rate, fluctuating during passage of wave, similar to fluctuations in orbital velocities shown by Figure 4-6.

k = calibration coefficient incorporating shear stress caused by orbital wave motions

$\Psi_{(t)}$ = instantaneous Shields parameter

Shields parameter Ψ includes a term for fluid shear stress τ_o, which initiates movement of sand with diameter D, and density ρ_s:

$$\Psi_{(t)} = \frac{\tau_{o(t)}}{(\rho_s - \rho_w)g\,D} \qquad (4\text{-}6)$$

where

$\tau_{o(t)}$ = instantaneous shear stress exerted near sea bottom by steady and oscillatory currents, fluctuating during passage of wave (Figure 4-6)

ρ_s = grain density

ρ_w = water density

D = grain diameter

Traction equations generally include a modification for shear stress, where an additional friction factor represents stresses imposed by oscillatory motions of waves. Madsen and Grant (1976, 77) and Dingler (1979) express bottom shear stress induced by mean and oscillatory currents as:

$$\tau_{o(t)} = \frac{1}{2} f_w \rho |u_t| u_t \qquad (4\text{-}7)$$

where

$\tau_{o(t)}$ = instantaneous shear stress exerted near sea bottom by oscillatory wave motion

f_w = wave friction coefficient (Jonnson, 1966)

u_t = instantaneous orbital velocity near sea bottom in x-y plane (Figure 3-3) whose components are given by Equations 3-22 and 3-23, and fluctuate during passage of wave (Figure 4-6)

Equation 4-7 is comparable to Equation 3-74, which is also an expression for wave-induced shear stress.

Use of shear stress modification is a popular means of simplifying the representation of oscillatory wave motions. Bagnold (1966), Bijker (1971), Ackers and White (1973), Swart (1974, 76), Bijker and others (1976), Madsen and Grant (1976, 77), Fleming and Hunt (1970), Lenhoff (1982), and Lee-Young and Sleath (1988) provide examples of traction-based equations that incorporate terms for representing wave-induced bed shear stress. Unfortunately, coefficient k in Equation 4-5 is a weak component of traction equations because calibration of k is not directly linked to readily measured parameters. Furthermore, calibration of k is required for each study area, limiting the use of traction equations for predicting sediment transport rates. Unlike coefficient K_3 in the wave-power equation (Equation 4-4, and Table 4-1), proponents of traction-based equations have not shown that coefficient k conforms to a predictable range. Galvin and Vitale (1976), Van de Graaff and Van Overeem (1979), Swart and Fleming (1980), Willis (1980), and Kamphuis and others (1986) provide further comparisons of traction-based and wave-power equations.

Energetics equations

Energetics equations relate sediment transport to energy supplied by waves to move bedload and suspended load (Bagnold,1963). Energetics equations incorporate general rules of physics for representing rates of sediment transport, where moving bedload, for example, is represented as a "block of moving sediment" (Figure 1-4B) and forces acting on the moving bedload are resolved using classic vector analysis. Sediment transport occurs when wave-induced forces overcome frictional forces. The energetics equations are hybrids between wave-power and traction equations because they include terms for wave-induced energy, shear stresses, and friction. Bagnold (1963) and Bowen (1980) expressed sediment transport rate Q in the form of bedload transport rate i_b as a function of bed slope, orbital velocities, friction, and the amount of energy available to do work, expressed as an "efficiency factor" ε_b :

$$i_b = \frac{\varepsilon_b\, C_D\,\, \rho\, u^3 |u|}{\tan\phi - \dfrac{u\beta}{|u|}} \tag{4-8}$$

where

i_b = transport rate of bedload
ε_b = efficiency of bed load transport (determined by experimentation)
C_D = drag coefficient (determined by experimentation)
ρ = fluid density
ϕ = critical friction angle or angle of repose
u = orbital velocity (fluid velocity at bottom) (Equation 3-26)
β = slope of beach

Similarly, they expressed suspended-load transport rate i_s as a function of fall velocity W of suspended sediment, orbital velocity u, and beach slope β :

$$i_s = \frac{\varepsilon_s\, C_D \rho\,\, u^3 |u|}{W - u\beta} \tag{4-9}$$

where

i_s = transport rate of suspended load
ε_s = efficiency of suspended load transport (determined by experimentation)
W = fall velocity of grains

Although Equations 4-5 through 4-9 seem straightforward, they represent only instantaneous quantities. Total transport rates must be determined by summing quantities over each wave period, for each passing wave. Obtaining these time-averaged expressions is difficult, inhibiting their use in computer procedures. Furthermore, efficiency factors ε_b and ε_s in Equations 4-8 and 4-9 are difficult to estimate from field data, making energetics equations difficult to calibrate and apply. Bailard (1981), Holman and Bowen (1982), and Bowen and Doering (1984) describe other procedures for representing transport based on modifications to Bagnold's energetics approach and Equations 4-8 and 4-9.

Empirical equations

While all transport equations are empirical, so called "empirical equations" are based more upon observational data than on physical principles. For example, Bascom (1951) observed that sand grain sizes are generally a function of beach slope, with steeper slopes characterized by larger grain sizes:

$$D = f(\tan\beta) \tag{4-10}$$

Hattori (1982) observed that onshore or offshore sand transport can be predicted statistically from wave steepness (ratio of wave height to wavelength) and proposed the following relationships:

If $[(H_0/L_0)\tan\beta]/(W/gT) > 0.5$, offshore transport occurs (4-11)

And if $[(H_0/L_0)\tan\beta]/(W/gT) < 0.5$, onshore transport occurs (4-12)

where

β = slope of beach
H_0 = wave height in deep water
L_0 = wavelength in deep water
W = fall velocity of mean grain size
g = gravitational constant
T = wave period

While useful, empirical equations add little to our understanding of the physical processes responsible for relationships between grain size, beach slope, wave climate, and direction of sediment transport, and their applications are generally limited to nearshore environments similar to those for which they were originally proposed.

WAVE's transport equation

Limitations imposed by WAVE's circulation module limit our choice of feasible transport equations. For example, WAVE cannot employ certain equations for onshore-offshore transport that rely on oscillatory currents because it employs depth-averaged and time-averaged currents that are unsuitable in these equations. As computer procedures improve for representing three-dimensional wave-induced currents, procedures for representing sediment transport can incorporate what are now only "theoretical" equations. Until then, our choice of transport equations is limited to those that employ time-averaged, depth-averaged currents.

To complicate our choice, no family of transport equations stands out as more or less empirical than the others. For example, wave-power equations include coefficient K (Equations 4-2, 4-3, and 4-4), traction equations include friction coefficients k and f_w (Equations 4-5 and 4-7) and energetics equations include efficiency and drag coefficients ε_b, ε_s, and C_D (Equations 4-8 and 4-9). However, the wave-power equation is derived from observations in the nearshore area where currents are assumed steady and uniform over the total depth, making it the most satisfactory choice for inclusion into WAVE. The wave-power equation (Equation 4-4) proposed by Komar and Inman (1970) is included in WAVE for the following reasons:

(1) It is widely used and is recommended by the Shore Protection Manual (U.S. Army CERC, 1977).

(2) Calibration of coefficient K is supported by field studies and lies within a relatively narrow range of values (Table 4-1).

(3) Parameters are readily available from field studies, with wave power calculated from wave heights and wave angles in the breaker zone.

(4) It is specifically devised for the surf zone, where transport and longshore currents are assumed to be steady, and is compatible with WAVE which calculates steady, depth-averaged and time-averaged currents.

(5) Oscillatory currents are not required.

(6) Multiple time steps are not required to represent fluctuations during the passing of individual waves or wave periods, allowing larger time steps that in turn allow simulations to span longer periods of time.

WAVE'S PROCEDURE FOR CALCULATING TRANSPORT RATES

WAVE uses the wave-power equation (Equation 4-4) by Komar and Inman (1970) to express volumetric transport rate Q in cubic meters per second. However, Komar and Inman express transport rate as an "immersed weight" after Bagnold (1963), who argues that transport rates should be expressed as immersed weights rather than as "volumetric rates" because sand grains may have various densities and may settle with different porosities. Consequently, WAVE calculates transport rates as immersed weights based on equations by Bagnold (1963) and Komar and Inman

97

(1970) and then converts them to volumetric rates to facilitate the the representation of the movement of sediment volumes between grid cells.

Immersed weight refers to the weight that grains have when submerged and is a function of the density of grains and surrounding water. By contrast, volumetric rates refer to the volume that moving grains would have if they were to settle and includes the pore space between grains, but does not consider their different densities. Conservation of mass is an important consideration and the use of immersed weights allows sediment transport to be expressed in terms of mass rather than volume. The "immersed-weight longshore transport rate" of sand I_l is generally expressed in (dynes/sec) or (newtons/sec) and is related to the volumetric transport rate Q (Equation 4-4) by:

$$I_l = Q \left(\rho_s - \rho_w \right) g\, a'$$ (4-13)

where

I_l = immersed-weight longshore transport rate
Q = volumetric longshore transport rate (generally in cubic meters per second)
ρ_s = grain density
ρ_w = water density
g = gravitational constant
a' = corrects bulk volume Q for pore space and is equivalent to $(1-\phi)$, where ϕ is the porosity of settled grains

The advantage of Equation (4-13) over Equation (4-4) is that I_l has the same units as wave energy flux P_l, which is either expressed as force per unit time, energy per unit length per unit time, or power per unit width, and is generally given in newtons per second, joules per meter per second, or watts per meter:

$$I_l = K\, P_l$$ (4-14)

where

K = coefficient of proportionality
P_l = wave-energy flux in longshore direction

Wave energy flux P_{sf} in the surf zone is equal to the product of wave energy and group velocity at the breaker zone:

$$P_{sf} = E_b\, C_b$$ (4-15)

where

E_b = wave energy per unit length of wave crest, or "wave-energy density" at breaker zone   (Equation 3-63)

C_b = group velocity, or "velocity of energy propagation" given by Equation 3-68, as evaluated in breaker zone

P_{sf} = total wave-energy flux in surf zone

The wave energy flux in the surf zone in the longshore direction is given by Komar and Inman (1970):

$$P_l = (E_b\, C_b)\, \cos \alpha_b \sin \alpha_b \qquad (4\text{-}16)$$

where

α_b = angle of incidence at breaker zone (Figure 4-8), denoted β in Equation 3-31

P_l = wave-energy flux in longshore direction

Combining Equations 4-14 and 4-16 gives an expression for the immersed-weight longshore transport rate, which is similar to Equation 4-4, except that the transport rate is expressed in terms of immersed weight rather than volume:

$$I_l = K\, (E_b\, C_b)\, \cos \alpha_b\, \sin \alpha_b \qquad (4\text{-}17)$$

where

K = coefficient of proportionality (Examples in Table 4-1)

Values given for K in Table 4-1 are appropriate only for Equation 4-17 involving immersed weights, whereas volumetric transport rate Q (Equation 4-4) must be converted to immersed weights (Equation 4-13) for values of K to be appropriate. WAVE uses Equation 4-17 to calculate immersed-weight longshore transport rates and then converts them to volumetric rates using Equation 4-13.

Calculating transport rates in a grid

WAVE's procedures for calculating transport rates employ the same parameters used in field studies which employ a simple "discharge equation" that includes terms for the velocity of moving bedload and cross-sectional area of the flow (Figure 4-9). This similarity of procedures allows direct calibration of WAVE's results to field data. For example, field studies measure wave power and transport rates to yield coefficient K (Equation 4-17), while WAVE simulates wave power and uses a constant K to yield transport rate. In field studies, the general steps involved in relating transport rates to wave energy and obtaining coefficient K are as follows:

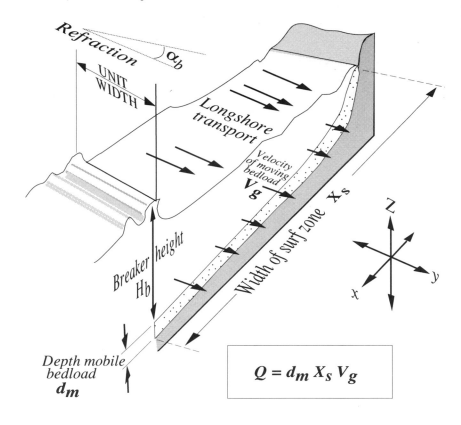

Figure 4-9 Variables in estimating longshore rate Q, which is a product of thickness of moving bedload d_m, width of surf zone X_s, and velocity of sand grains V_g.

<u>Step 1:</u> Obtain volumetric transport rate in the surf zone with Equation 4-18 by measuring thickness of moving bedload, width of the surf zone, and velocity of sand grains :

$$Q = d_m \, X_s \, V_g \qquad (4\text{-}18)$$

where

Q = volumetric longshore transport rate (Figure 4-9)
d_m = depth of mobile bed or "depth of disturbance" (King,1951; Madsen ,1989)
X_s = width of surf zone between breaker zone and maximum swash
V_g = velocity of grains moving as bedload and suspended load
        (Komar and Inman (1970) provide examples of d_m , X_s, and V_g)

Step 2: Express Q as an immersed weight I_l with Equation 4-13.

Step 3: Obtain longshore wave energy flux P_l with Equation 4-16.

Step 4: Plot immersed weight transport rate I_l versus wave energy flux P_l (similar to Figure 4-2), to obtain coefficient K (Equation 4-14).

WAVE's steps for calculating sediment transport rate Q proceed in reverse order, beginning with an estimate of coefficient K, and ending with an estimate for the volumetric transport rate Q in cubic meters per second, as follows:

Step 1: Estimate K from field data (Table 4-1 or Figure 4-2).

Step 2: Obtain longshore wave energy flux P_l (Equation 4-16) by simulating wave heights and incident angles at breaker zone. WAVE's algorithm proceeds up each column of the grid searching for the first cell whose wave height exceeds the maximum breaker height, signifying that waves break and establishing the location of the breaker zone (Figure 4-10).

Step 3: Once the breaker zone is located, the immersed weight transport rate I_l is obtained from Equations 4-14 and 4-16 evaluated at the breaker zone.

Step 4: Express immersed weight transport rate I_l as volumetric transport rate Q (Equation 4-13).

The volumetric transport rate Q (Equation 4-13) is the total transport rate over the full width of the surf zone (Figure 4-11). It can be geographically distributed evenly among cells within the surf zone, or unevenly to mimic more closely the distribution of transport rates observed on actual beaches (Figure 4-12). Plots of transport rates within the surf zone generally conform to a skewed bell-shaped curve, with the highest rates near the breaker zone where longshore currents are generally greatest (Ingle, 1966, p. 99; Komar, 1971; and Kraus and others, 1988). However, rates may not correlate directly with velocities of longshore currents because bed shear stress, orbital velocities, and other factors create additional turbulence that may locally increase rates of sediment transport (Figure 4-12). WAVE calculates maximum orbital velocities (Equation 3-26) in the breaker and surf zone to conform to profiles similar to Figures 3-11 and 4-12, allowing Q to be distributed within the surf zone as follows:

$$Q_{i,j} = Q_{tot} \ (U_{max_{i,j}} / U_{tot}) \tag{4-19}$$

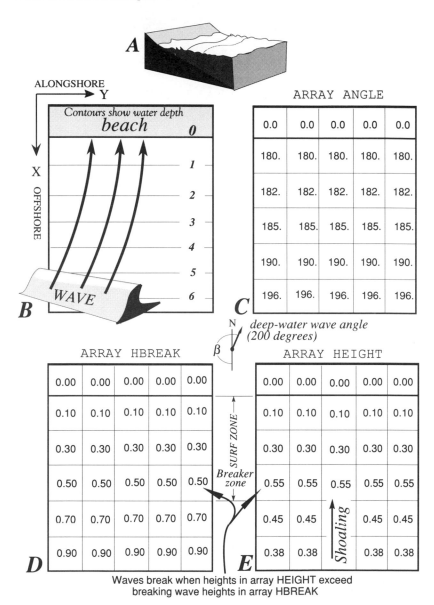

Figure 4-10 Schematic representation showing how WAVE establishes location of breaker zone. Coordinate scheme for arrays is shown in Figure 3-13. WAVE finds location of breaker zone and determines rates of sediment transport within surf zone extending between breaker zone and beach. (A) Three-dimensional representation of beach. (B) Contour map of water depth. WAVE propagates wave crests over grid and provides (C) array ANGLE of wave angles in degrees, (D) array HBREAK of breaking wave heights in meters, and (E) array HEIGHT of shoaling wave heights in meters. Waves break when heights in array HEIGHT exceed those in array HBREAK.

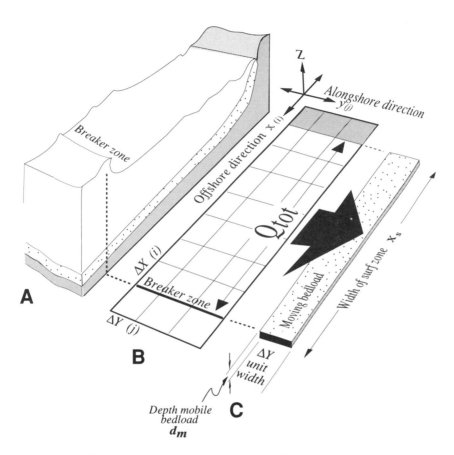

Figure 4-11 Diagram showing how total volumetric longshore transport rate Q_{tot} is calculated by WAVE, which incorporates procedures for distributing total bedload among individual cells within surf zone based on profiles of natural beaches (Figures 4-13 and 4-14): (A) Segment defining surf zone extending from beach to breaker zone. (B) Grid representing segment of surf zone in A where sediment transport rates are calculated within each cell. (C) Rate of sediment transport Q_{tot} within surf zone is obtained from wave-power equation (Equation 4-17), which assumes that rate over entire width of surf zone X_s applies to moving bedload of uniform thickness.

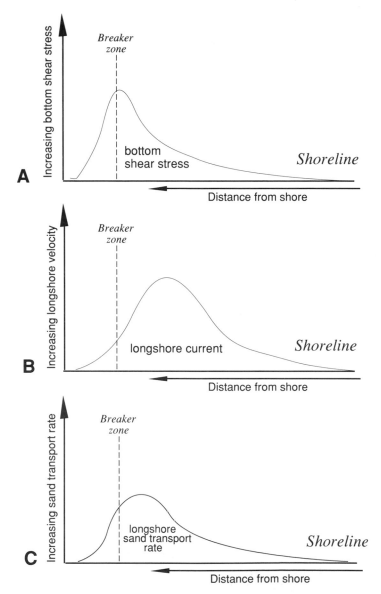

Figure 4-12 Hypothetical profiles perpendicular to shore showing variations of bottom shear stress, longshore velocity, and sand transport rate relative to breaker zone and shore on typical beach: (A) Bottom shear stress is generally greatest in breaker zone. (B) Longshore current velocity is generally greatest between breaker zone and shoreline. (C) Longshore transport rate is related both to shear stresses and longshore currents, and is generally greatest between breaker zone and place where longshore currents are greatest. Modified from Longuet-Higgins (1970) and Komar (1971).

where

i, j = subscripts corresponding to x and y axes, shown by Figure 4-11
Q_{tot} = volume transport rate in each column of grid, over total width of surf zone
$Q_{i,j}$ = volume transport rate in cell i, j
$U_{max_{i,j}}$ = maximum orbital velocity at grid location i, j (Equation 3-26)
U_{tot} = sum of maximum orbital velocities in surf zone

Equation 4-19 allows reasonable representation of variable transport rates in the surf zone (Figure 4-13). Where beaches are irregular, WAVE simulates waves that break at various water depths and grid locations. Similarly, Equation 4-19 allows rates of sediment transport to differ between adjacent grid cells because wave energy at the breaker zone varies from column to column. Thus, transport rates along irregular shorelines continually change as topography, wave climate, and location of the breaker

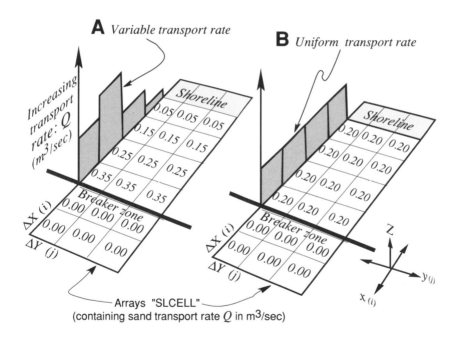

Figure 4-13 Diagram contrasting two schemes for distributing transport rate Q among cells within surf zone. Numbers in cells represent hypothetical transport rates in cubic meters per second: (A) Variable transport rate in which geographic distribution of sediment transport rates among individual cells are highest where bottom shear stress is highest. (B) Uniform transport rate assumes that rates are uniform within surf zone. Rates are stored in array SLCELL.

105

zone change (Figure 4-14). Once transport rates are calculated for each grid cell in the surf zone (Figures 4-13 and 4-14), WAVE can erode, transport, and deposit sediment based on rates in each cell. Chapter 5 describes accounting procedures for recording sediment movement to and from cells.

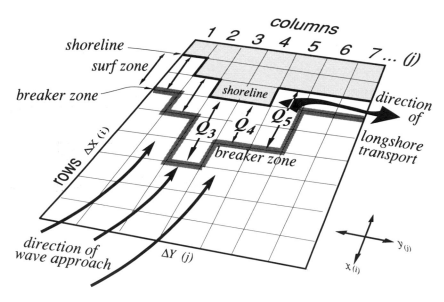

Figure 4-14 Geographic distribution of total sediment transport rate Q apportioned among columns of array SLCELL representing area of beach. Variations in depth and wave climate at grid cells cause sediment transport rates in surf zone to vary alongshore as positions and wave energy of breaker zone change. Variables Q_3, Q_4 and Q_5 representing total transport rates in columns 3, 4, and 5 are functions of incoming wave energy within columns 3, 4 and 5. Sand within surf zone generally moves parallel to shoreline.

SUMMARY

WAVE employs the wave-power equation proposed by Komar and Inman (1970) to estimate rates of longshore transport. It is best applied in the surf zone where currents are assumed to be steady. Its selection is dictated by constraints imposed in WAVECIRC, which calculates steady rather than oscillating currents. While traction and energetics equations may be better for representing onshore-offshore transport, their solution requires "unaveraged" oscillatory currents, and are thus not appropriate for WAVE. The wave-power equation is easy to employ and includes familiar terms such as wave height, wave angle, and wave energy, which are readily calculated by WAVECIRC and easily compared with parameters reported from field studies.

CHAPTER **5**

Simulating Erosion, Transport, and Deposition

Once transport rates are calculated, WAVE employs procedures to represent erosion, sorting, and transport of sand grains of variable diameter and density. WAVE includes accounting schemes for representing areas, masses, and volumes of sediment in a three-dimensional record that can change through time in response to longshore transport. This chapter describes WAVE's procedures for representing longshore transport (Figure 5-1) and documents their performance with a series of experiments.

REPRESENTATION OF SPACE AND TIME

WAVE employs arrays that represent area or space divided into discrete cells. Harbaugh and Bonham-Carter (1970) provide an overview of the use of arrays in geological applications. For example, WAVE employs a two-dimensional array (Figure 5-1B) to represent sediment transport rates along a coastline, where each cell contains a specific transport rate in cubic meters per second. Values in arrays are referenced by row index i and column index j that correspond to x and y coordinates. Similarly, WAVE uses a two-dimensional array to store topographic elevations (Figure 5-2A) and a four-dimensional array (Figures 5-2 and 5-3) to store the location, age, thickness, and composition of sediment cells that represent sedimentary deposits.

Ages of sediment cells
Each "sediment cell" deposited by WAVE records the volume of sediment deposited in a specific interval of time. Limited computer memory prohibits the representation of every sedimentary layer that has contributed to a sequence of beds

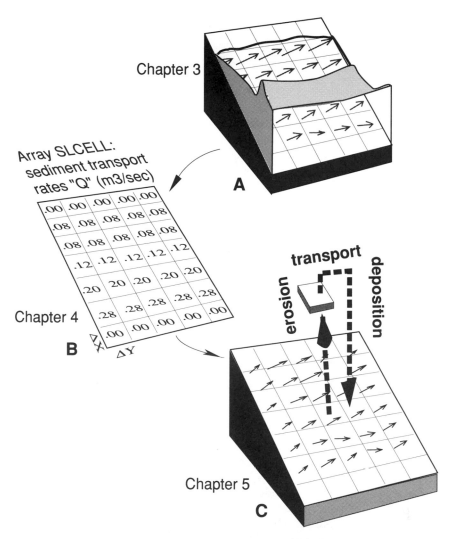

Figure 5-1 WAVE's three main steps in computing sediment transport by waves in surf zone: (A) Compute shoaling and refraction angles of waves, location of breaker zone, and currents (arrows) within surf zone. (B) Compute sediment transport rates in m^3/sec for grid cells within surf zone. (C) Calculate volumes of sediment eroded from grid cells and transported to adjacent cells.

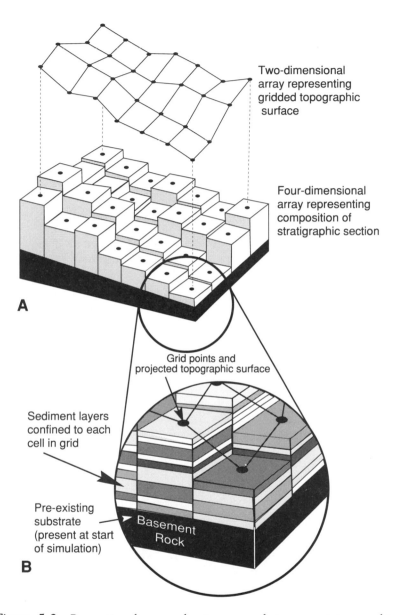

Figure 5-2 Perspective diagrams showing arrays that represent topographic surfaces and sediment layers within cells: (A) Sedimentary record is contained in arrays, with topographic elevations defined at center of grid cells in two-dimensional array, and location, composition, thickness, and location of sediment layers contained in four-dimensional array. (B) Enlargement showing layers of varying thickness and composition represented by separate "sediment cells". Basement denotes substrate that cannot be eroded.

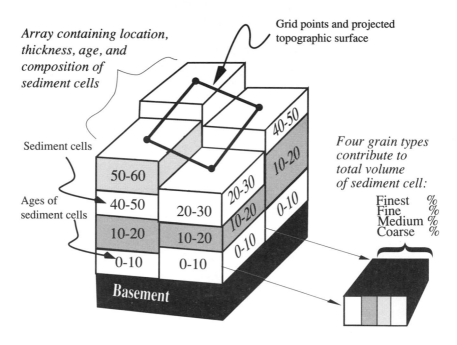

Figure 5-3 WAVE's representation of sediment layers in which four-dimensional array stores location, thickness, age, and composition of sediment cells. "Age" of cells refers to time interval during which sediment is deposited, where age of "0-10" contains sediment deposited between 0 and 10 years after the start of simulation.

during an experiment, and therefore the number of sediment cells represented during a simulation must be limited. For example, WAVE may require 25,000 iterations to simulate 25 years, during which 25,000 layers of sediment are deposited, but each layer cannot be stored. Instead, sediment cells are employed to represent an aggregate of sediment layers deposited during a fixed interval of time, and cells are "closed" (Figure 5-4) at a regular interval prescribed as input. For example, cells might be closed every year so that only 25 cells are required in each grid location to represent 25 years. Thus, the sediment cells may represent the average composition of several lesser sedimentary layers, although the thicknesses of sediment cells can still vary (Figure 5-5C). Alternatively, some computer procedures employ a scheme where cells are closed once they attain a certain thickness (Figure 5-5B), and while these procedures record vertical variations in sediment composition, variations in bed thicknesses are lost.

Composition of sediment cells

A four-dimensional array is required to record geographic location, ages, and composition of cells (Figure 5-3), with up to four grain types represented in each cell.

110

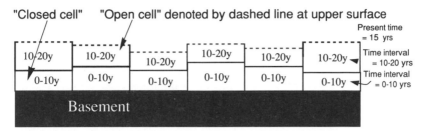

Figure 5-4 Schematic cross section defining "open" versus "closed" sediment cells. Cells that receive sediment during current time interval are open (dashed line on top), whereas older cells beneath are closed (continuous line on top). Here, cells labeled "0-10y" contain sediment deposited during first 10 years of simulated time and are now closed, whereas cells labeled "10-20y" contain sediment being deposited during present interval extending from 10 to 20 years.

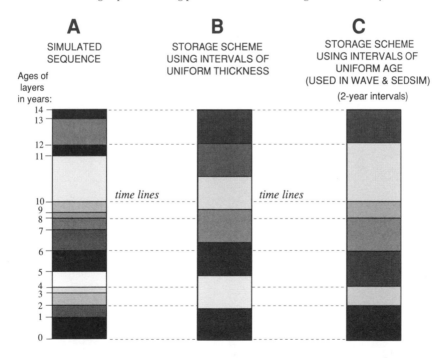

Figure 5-5 Columns of sediment layers showing alternative methods for representing simulated stratigraphic sequences where shading denotes variations in composition:(A) Simulated sequence of sedimentary layers deposited during varying intervals of time. (B) Representation of simulated sequence shown in A with layers of equal thickness in which composition of each layer is average of composition of thinner layers within it. (C) Procedure used by WAVE and SEDSIM where sequence of layers in A is represented by sediment cells spanning uniform intervals of time, thereby allowing sediment cells to vary both in thickness and composition.

111

In this way, a three-dimensional representation of the stratigraphic record including differences in grain size within layers can be preserved and displayed as sediment moves into and out of cells (Figure 5-6). While some computer programs write results to external files or tapes to reduce computer memory demands, WAVE and SEDSIM cannot employ this simplification because sediment deposited during previous time intervals may be eroded later as hundreds or thousands of iterations are performed to represent the passage of time (Figure 5-7). Thus, a four-dimensional record of ages and compositions of cells must be continuously maintained in memory to record depositional as well as erosional events.

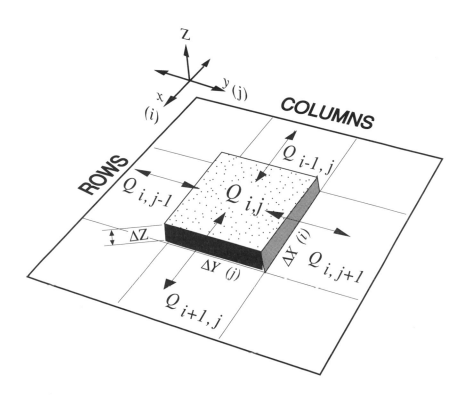

Figure 5-6 WAVE's procedure for transporting sediment across cell boundaries. Transport rates Q into and out of cell i, j contribute to volume change in cell. Because Δx and Δy are fixed, change in volume is proportional to change in thickness Δz.

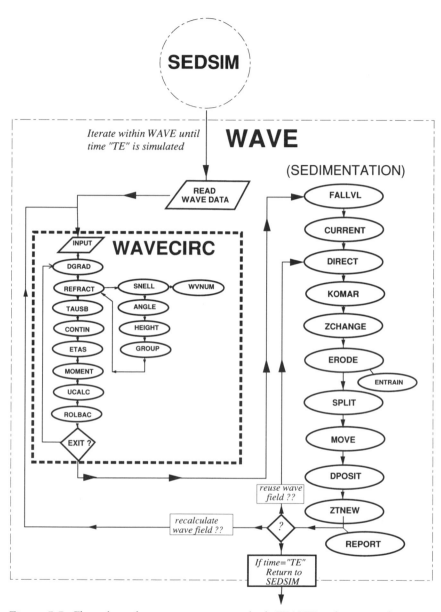

Figure 5-7 Flow chart showing sequence in which WAVE's subroutines for wave-induced circulation (WAVECIRC) and sediment transport are utilized. WAVECIRC forms separate module with its own subroutines. SEDSIM calls WAVE at time interval TE, specified in years or fraction of years. Wave-induced currents and sediment transport are calculated for interval TE, before control is returned to SEDSIM.

CONTINUITY EQUATION FOR SEDIMENT TRANSPORT

WAVE transports sediment by moving volumes of sediment to and from cells (Figure 5-6) according to a sediment-continuity equation where changes in density across cell boundaries during a time interval are given by:

$$\frac{\partial \rho}{\partial t} = \frac{\partial \rho v_x}{\partial x} + \frac{\partial \rho v_y}{\partial y}$$

(5-1)

where

ρ = density of sediment
t = time
v_x = velocity of sediment in x coordinate direction
v_y = velocity of sediment in y direction
x,y = coordinate axes of grid scheme (Figure 3-13)

Similarly, Equation 5-1 can be expressed in terms of sediment volume:

$$\frac{\partial Vol}{\partial t} = \frac{\partial Vol(v_x)}{\partial x} + \frac{\partial Vol(v_y)}{\partial y}$$

(5-2)

where

Vol = total volume of sediment moving to or from cell
v_x = velocity of sediment moving into cell along x axis
v_y = velocity of sediment moving into cell along y axis

However, it is more convenient to express changes in volume as a function of sediment transport rates Q (Figure 5-6), as provided by Equation 5-3, where change in a cell's volume is proportional to change in the rate of sediment transported to or from the cell:

$$\frac{\partial Vol}{\partial t} = \partial Q_x + \partial Q_y$$

(5-3)

where

Q_x, Q_y = volumetric rate of sediment transport with respect to x and y coordinates (Figure 5-6)

Erosion or deposition in a cell changes thickness Δz, so that a change in cell volume is the product of cell area $\Delta x \Delta y$ and the change in thickness Δz. The general continuity equation governing changes in thickness during time t is:

$$\frac{\partial z}{\partial t} = \frac{\partial Q}{\Delta x \Delta y} \tag{5-4}$$

where

Q = total volumetric rate of sediment moving into cell from four neighboring cells (Figure 5-6)
$\Delta x \Delta y$ = area of cell in x and y coordinates (Figure 5-6)

Thus, a change in thickness of a cell also changes topographic elevation z, and is related to the rate of sediment transport Q into or out of the cell by Equation 5-4.

REPRESENTING SORTING WITH FOUR GRAIN TYPES

The sediment-continuity equation (Equation 5-4) is too simple to represent sorting because it does not consider different grain types that contribute to the total volume of sediment moving to or from cells. WAVE simulates sorting by adjusting the relative contribution of various grain types to the total moving volume, where the relative contribution of each grain type is a function of grain diameter, density, and availability. Thus, total sediment volume Vol moving to and from cells is the sum of the contributions of volumes of each grain type Vol_{ks} :

$$Vol = \sum_{ks} Vol_{ks} \tag{5-5}$$

where

Vol = total volume of sediment moving to or from a cell
Vol_{ks} = volume of grain type ks
ks = grain types 1, 2, 3, 4

WAVE allows a maximum of four grain types to be eroded and moved, where the diameter and density of each grain type are provided as input. Sediment in WAVE can therefore be represented by a mixture of four grain types, where the volumetric contribution of each grain type Vol_{ks} to the total eroded or moving volume of sediment Vol, is a function of several parameters:

$$Vol_{ks} = f(\rho_s, \rho_w, D, W, \mu, g, \tau_b) \tag{5-6}$$

115

where

ρ_s	= density of sediment	μ	= fluid viscosity
ρ_w	= density of water	g	= gravitational acceleration
D	= average grain diameter	τ_b	= bottom shear stress
W	= fall velocity		

Parameters in Equation 5-6 are used to estimate the "transport efficiency" ε_{ks} for each grain type, which is the relative contribution of each grain type as a percentage of the total volume *Vol* moving to and from cells. Calculation of transport efficiencies allows WAVE to represent sorting, reworking, and the movement of each grain type present. For example, grains that are slow to be eroded or moved have low transport efficiencies and contribute less to the total volume of sediment moving to and from a cell than more easily moved grains which have high transport efficiencies and contribute more to the total volume. Thus, Equation 5-5 can be expressed in terms of transport efficiencies of available grain types:

$$Vol = \sum_{ks} Vol \; \varepsilon_{ks} \qquad (5\text{-}7)$$

where

ε_{ks} = transport efficiency as a percentage where

$$\sum_{ks} \varepsilon_{ks} = 100 \, \% \qquad (5\text{-}8)$$

Therefore, the volumetric contribution Vol_{ks} of each grain type to the total volume *Vol* moving to or from a cell can be calculated from Equation 5-9 if transport efficiencies of individual grain types are known:

$$Vol_{ks} = Vol \cdot \varepsilon_{ks} \qquad (5\text{-}9)$$

CALCULATING TRANSPORT EFFICIENCIES

WAVE employs Equation 5-9 to calculate the volumetric contribution of each of four grain types Vol_{ks} to the total volume of sediment *Vol* eroded from a cell. WAVE calculates each volumetric contribution by calculating transport efficiencies ε_{ks} as functions of grain diameter and density, thus representing sorting by eroding varying amounts of each of the four grain types. WAVE, however, cannot represent all mechanisms involved in sorting. For example, WAVE does not simulate "transport sorting" (Komar, 1989; Figure 5-8) where some grains move farther or faster than

A. Selective movement

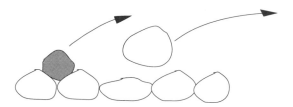

B. Settling equivalence

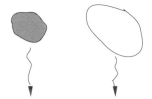

C. Transport sorting

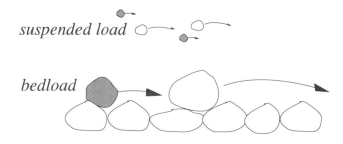

suspended load

bedload

Figure 5-8 Diagram modified from Komar (1989) showing transport (arrows) and sorting of grains where grain diameters and densities (denser grains are shaded) have major influence: (A) Selective movement may cause larger grains to move before smaller denser grains. (B) Grains with different diameters may settle at same rates because of differences in density. (C) Smaller grains moving as suspended load may be transported more slowly than larger grains moving as bedload, although once suspended, smaller grains may move farther.

others, because grain types eroded during a single time step can be transported only a fixed distance Δx to immediately adjacent cells (Figure 5-6). However, "selective movement" of sediment (Komar, 1989; Figure 5-8) is represented in WAVE by allowing the relative contributions of grain types moving from each cell to vary according to their transport efficiencies, which are themselves related to grain size and density.

Physical principles governing selective movement of sediment by steady currents also apply to movement by oscillatory currents. For steady currents, Shields (1936) demonstrated that initiation of sediment movement is a function of Shields parameter Ψ and particle Reynolds number R_e (Equations 5-10, 5-11, and 5-12, and Figure 5-9):

$$Q = f(\Psi, R_e) \tag{5-10}$$

$$\Psi = \frac{\tau_c}{(\rho_s - \rho_f)gD} \tag{5-11}$$

$$R_e = \frac{\rho U^* D}{\mu} \tag{5-12}$$

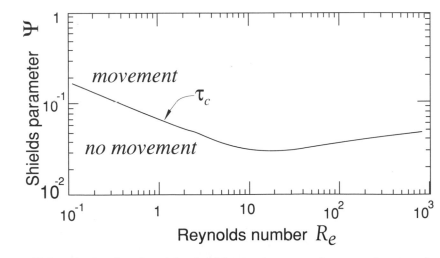

Figure 5-9 Log-log plot of threshold for movement τ_c of grains as function of Shields parameter Ψ with respect to Reynolds number R_e. Area above plot represents conditions under which sediment is moved. After Shields (1936) and Miller and others (1977).

where

Ψ = Shields' parameter
R_e = Reynolds number
τ_c = critical boundary shear stress, or threshold stress. Equation 4-7 provides. general expression for wave-induced stress required to move grains
ρ_s = grain density
ρ_f = fluid density
g = gravitational acceleration
μ = fluid viscosity
D = grain diameter
U^* = shear velocity = $(\tau_c/\rho_f)^{1/2}$

Of variables needed for Equations 5-11 and 5-12, only threshold shear stress τ_c is unknown and it can be determined from Figure 5-9 by solving Equations 5-11 and 5-12 simultaneously. However, Miller and others (1977) and Komar (1989) provide a more useful relationship for determining τ_c, where effects of mixed grain types are represented by a term for bed roughness that includes pivot angle Φ:

$$\tau_c = 0.00515(\rho_s - \rho_f)gD^{0.568}\tan\Phi \qquad (5\text{-}13)$$

where
units shown in variable definitions are particular to Equation 5-13:
τ_c = threshold stress in dynes per cm^2
ρ_s = grain density in gm per cm^3
ρ_f = fluid density in gm per cm^3
g = gravitational constant expressed as 981.0 cm per sec^2
D = grain diameter in mm
Φ = pivot angle (Figure 5-11)
$\tan\Phi$ = tangent of pivot angle, providing an expression for bed roughness ($\tan\Phi$ is approximately 0.6 for grains of uniform diameter)

Figure 5-10 is a graph of Equation 5-13 for threshold stress τ_c versus grain diameter D that pertains to uniform currents and populations of grains having uniform diameters and densities. Figure 5-10 shows that larger shear stresses are required to move larger or denser populations of grains, whereas small shear stresses are required to move smaller or less dense grains.

Where mixtures of grains are present, the correlation between threshold shear stress and grain diameter may diverge from trends in Figure 5-10 (Komar and Clemens, 1986), and additional terms such as the pivot angle Φ (Equation 5-14, Figure 5-11) are employed to describe movement of grains relative to others of different size or density. Pivot angle Φ helps quantify differences between different populations of grain types, allowing WAVE to represent selective entrainment of grains in nonhomogeneous

119

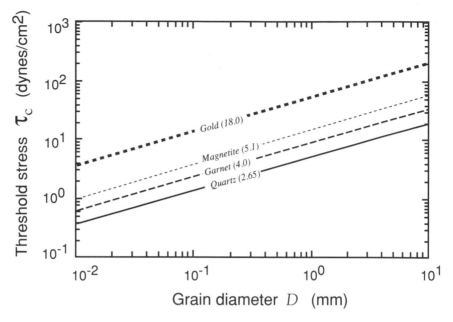

Figure 5-10 Log-log plot of threshold shear stress τ_c versus grain diameter D calculated with Equation 5-13, for quartz, garnet, magnetite, and gold. Numbers in parentheses are densities in gm/cm^3. Based on Miller and others (1977).

mixtures. In a general sense, pivot angle is a measure of bed roughness. Its use in governing selective entrainment is described by Miller and Byrne (1966) and Slingerland and Smith (1986), and by Komar (1989), who provides the following equation for pivot angle:

$$\Phi = e \left(\frac{D}{K} \right)^{-f} \tag{5-14}$$

where

D = diameter of grain being entrained
K = diameter of grains over which grain rolls or pivots
e, f = calibration coefficients which are functions of angularity

Disregarding calibration coefficients, pivot angle Φ is proportional to the ratio of average diameters K of grains versus the diameter of a moving grain D, allowing Equation 5-14 to be expressed as:

$$\Phi \approx \frac{K}{D} \tag{5-15}$$

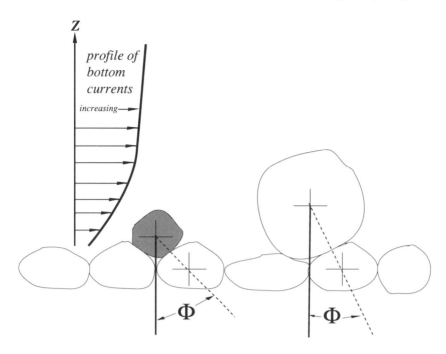

Figure 5-11 Cross section at water-sediment interface showing relationship between pivot angle Φ, particle size, and current velocities. Pivot angle is defined by dashed lines through centers of grains with respect to vertical. Larger grains have smaller pivot angles and protrude further above surface, subjecting them to higher velocities and enabling them to be preferentially transported. Modified from Komar (1989).

Where grains are of uniform size, Komar (1989) suggests that the pivot angle is approximately 30 degrees:

$$\Phi \approx 30 \ \text{degrees} \tag{5-16}$$

Therefore, for grains of uniform size, tanΦ in Equation 5-13 is assumed to be approximately equal to the tangent of thirty degrees, so that tanΦ can be rewritten as:

$$\tan \Phi \approx 0.6 \tag{5-17}$$

Equations 5-15 and 5-16 suggest that if a grain's diameter D is large relative to the average diameter of grains K, its pivot angle will be proportionally greater than 30 degrees. If its diameter is small relative to K, the pivot angle will be proportionally less than 30 degrees. Thus, pivot angle can be expressed as being proportionally greater or smaller than thirty degrees, depending on the ratio of K/D (Equation 5-18), where K/D provides a measure of bed roughness which helps quantify heterogeneity of a mixture of grain types:

$$\Phi \approx \left(\frac{K}{D}\right) \cdot 30^{0} \qquad\qquad (5\text{-}18)$$

where

D = diameter of grain being entrained
K = average diameter of grains in mixture

WAVE calculates pivot angle Φ with Equation 5-18, with diameters of different grain types D provided by the user, and average diameter K calculated from mixtures of four grain types in each cell (Figure 5-3).

Once pivot angle Φ is obtained, threshold shear stress τ_c is obtained with Equation 5-13, thereby providing a means of estimating critical shear stress for different types of grains in a mixture. Incorporation of pivot angle in Equation 5-13 may permit larger grains to move more readily than smaller grains. For example, grains larger than their neighbors have small pivot angles, and those with smaller diameters have larger pivot angles (Figure 5-11). Grains with small pivot angles tend to move more readily even though they may be larger than other grains, because they protrude higher above the surface and are more likely to be moved by currents.

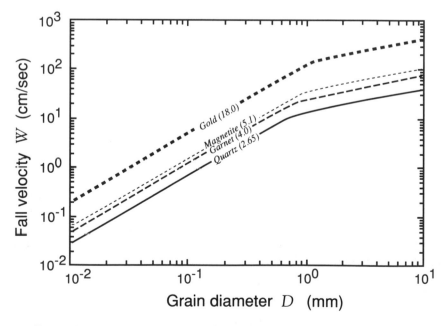

Figure 5-12 Log-log plot of fall velocity W versus grain diameter D calculated with Equation 5-26 for quartz, garnet, magnetite, and gold, whose densities are given in gm/cm^3.

Threshold shear stress τ_c provided by Equation 5-13 is important in calculating transport efficiencies ε_{ks} used by WAVE to determine the relative contribution of each grain type to the sediment volume moving to and from cells (Equation 5-9). Transport efficiencies of different grain types ε_{ks} are inversely proportional to threshold shear stress, where grains with large threshold shear stress have lower transport efficiencies and are less likely to move than those having small threshold stress:

$$\varepsilon_{ks} \approx \frac{1}{\tau_c} \qquad (5\text{-}19)$$

Fall velocity W of grains provides another parameter for estimating transport efficiency. Figure 5-12 shows a plot of fall velocity versus grain diameter for gold, magnetite, garnet, and quartz, where grains with lesser densities have lower fall velocities, suggesting that they would be transported farther and have higher transport efficiencies than denser grains. Thus, transport efficiencies are also inversely proportional to fall velocities:

$$\varepsilon_{ks} \approx \frac{1}{W} \qquad (5\text{-}20)$$

Steidtmann (1982) and Slingerland and Smith (1986) used the ratio of shear velocity U^* (Equation 5-12) to fall velocity W for predicting movement of mixtures of grains. They show that grain velocities are proportional to shear velocities and inversely proportional to fall velocities, suggesting that transport efficiencies are generally proportional to shear velocities:

$$\varepsilon_{ks} \approx U^* \qquad (5\text{-}21)$$

Combining Equations 5-19, 5-20, and 5-21 gives the approximation:

$$\varepsilon_{ks} \approx \frac{U^*}{W \, \tau_c} \qquad (5\text{-}22)$$

where transport efficiency is proportional to shear velocity and inversely proportional to fall velocity and threshold shear stress. Slingerland and Smith (1986) suggests that shear or friction velocity U^* is an expression for turbulence and can be estimated from near-bottom currents or orbital currents calculated in Equations (3-22) and (3-23). However, WAVE uses a simpler formula by defining coefficient C_f to represent turbulence. Substitution of C_f in Equation (5-22) gives:

$$\varepsilon_{ks} \approx \frac{C_f}{W \, \tau_c} \qquad (5\text{-}23)$$

where

C_f = calibration coefficient that includes effects of turbulence

WAVE assumes that C_f is equal to 1.0, but further calibration of C_f can yield transport efficiencies that correlate more closely to measured grain velocities on actual beaches.

Estimates of fall velocity W for use in Equation 5-23 are also complicated if mixtures of grains are involved. If grains are spherical and fall in low concentrations through still water, their fall velocities can be obtained by equating weight and drag forces (Tetzlaff and Harbaugh, 1989):

$$W = \left(\frac{\rho_s - \rho_f}{\rho_f} \frac{4}{3} \frac{gD}{C_d} \right)^{1/2} \tag{5-24}$$

where

C_d = drag-coefficient

Observed values of drag-coefficient C_d as a function of Reynolds number are graphed in Figure 5-13. WAVE calculates fall velocities by using drag-coefficient values in Figure 5-13 after the method of Tetzlaff and Harbaugh (1989). Within the range of Reynolds numbers where drag coefficients remain constant at 0.55, Equation 5-24

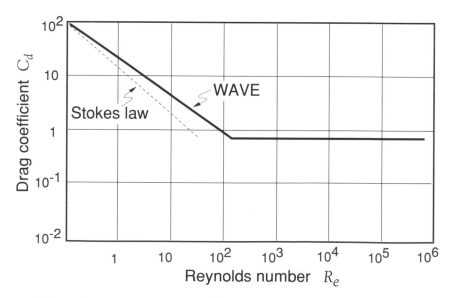

Figure 5-13 Function used by WAVE (heavy line) relating drag coefficients C_d to Reynolds number R_e. Function defined by Stokes law is shown by sloping dashed line, where $C_d = 24/R_e$ (Slingerland and Smith, 1986). Modified from Tetzlaff and Harbaugh (1989).

yields fall velocity. Where the drag coefficient is represented by the inclined straight line in Figure 5-13, fall velocity is:

$$W = \left(\frac{\rho_s - \rho_f}{\rho_f} \frac{4g}{90} \right)^{4/5} v^{3/5} D^{7/5} \tag{5-25}$$

where

v = kinematic viscosity of water

Tetzlaff and Harbaugh show that Equations 5-24 and 5-25 can be combined for calculating fall velocity:

$$W = MIN(4.88\gamma^{1/2} D^{1/2}, 1895\gamma^{4/5} D^{7/5}) \tag{5-26}$$

where

$\gamma = (\rho_s - \rho_f)/\rho_f$

Fall velocities in Figure 5-12 were obtained with Equation 5-26 and conform closely with those presented by Komar and Clemens (1986), Slingerland and Smith (1986), and Komar (1989).

WAVE uses fall velocities to calculate transport efficiencies ε_{ks} (Equation 5-23), which are inversely proportional to threshold shear stress τ_c and fall velocities W of each grain type present in a cell. While Equation 5-23 requires calibration of coefficient C_f, it does provide a theoretical method for calculating the relative contributions of different grain types during erosion and transport of sediment.

ORGANIZATION WAVE'S SEDIMENT TRANSPORT MODULE

The organization of WAVE's transport module is outlined in Figure 5-7 and Table 5-1. WAVE consists of the circulation module WAVECIRC described in Chapter 3 and a collection of subroutines for representing sediment transport, collectively referred to as WAVE's sediment transport module. WAVE begins by calling WAVECIRC, which calculates various parameters describing the hydrodynamics of waves. Next, fall velocities of each of four grain types are calculated, and directions of sediment transport are determined from CURRENT and DIRECT. KOMAR then calculates rates of sand transport based on wave-induced currents calculated within WAVECIRC and employs ZCHANGE to calculate the change in elevation of cells according to Equation 5-4. Next, WAVE employs ERODE and ENTRAIN to erode grain types in different amounts as a function of their density and

Table 5-1 Subroutines in WAVE. WAVECIRC forms separate module whose subroutines are described in Table 3-1 and shown in Figure 5-7.

Subroutine	Purpose
READWV	Reads input data file for WAVE
WAVECIRC	Wave-circulation model described in Chapter 3
FALLVL	Calculates fall velocities of each of four grain types
CURRENT	Determines directions of longshore currents provided by WAVECIRC
DIRECT	Determines direction of sediment transport as function of incoming wave angle and local topographic slope
KOMAR	Calculates sediment transport rates by method described in Chapter 4
ZCHANGE	Determines time-step and calculates topographic change using continuity equation
ERODE	Erodes sediment from cell and accounts for changes in composition
SPLIT	Expresses volumetric transport in x and y directions
MOVE	Moves eroded sediment to adjacent cells
DPOSIT	Deposits sediment and readjusts composition of sediment layers
ZTNEW	Updates array containing topographic surface
REPORT	Writes output file containing transport rates

diameter. Once volumes of sediment are eroded, they are moved to immediately neighboring grid cells by MOVE, and arrays are adjusted in DPOSIT and ZTNEW to account for the increase or decrease of sediment in each cell.

SIMULATING EROSION

Erosion is the most involved procedure within WAVE's sediment transport module because WAVE must represent the movement of mixtures of different grain types, with each type behaving differently according to its grain diameter and density.

126

WAVE begins its representation of erosion by first calculating the volume of sediment that could be removed (assuming it is available) from each cell based on transport rates in each cell (Equations 5-3). Erosion occurs by decreasing each cell's thickness according to the sediment continuity equation (Equation 5-4), where the change in a cell's elevation Δz during time step t_{step} depends on the rates of sediment transport into or out of the cell, whose area is Δx by Δy (Figure 5-6). Rearranging Equation 5-4 provides an expression for the volumetric change in a cell given as a change in elevation Δz :

$$\Delta z = \frac{Q}{\Delta x \Delta y} \, t_{step} \qquad (5\text{-}27)$$

where

t_{step} = time step over which sedimentation occurs, represents the time in seconds represented by each call to WAVE's sediment transport module
Q = transport rate of total sediment volume into or out of cell
Δz = change in surface elevation
$\Delta y, \Delta x$ = cell dimensions along x and y axes
x, y, z = coordinate directions (Figure 4-16)

Each iteration involving Equation 5-27 spans a time step t_{step} ranging from a few seconds to a few hours, so that during each call to WAVE, many iterations may be required to represent sediment transport over a longer interval of time.

Time step t_{step} determines the thickness of sediment eroded from a cell during each iteration of WAVE's sediment-transport module (Figure 5-7). If the time step is not chosen carefully, too much erosion can occur where high transport rates remove excessive volumes of sediment, creating unrealistic topographic depressions. To avoid this problem, WAVE uses a self-adjusting time step, where the maximum depth of erosion z_{max} during an iteration involving Equation 5-27 is controlled by calculating an optimum time step t_{step} that is small enough that erosion occurring during an iteration is less than the maximum depth of the mobile bedload z_{max} provided by variable DMOBILE as input (Appendix C).

Subroutine ZCHANGE governs the calculation of t_{step} by first locating the largest sediment transport rate Q in array SLCELL (Figure 5-1B), which becomes Q_{max} in Equation 5-28. The maximum amount of time represented by one iteration, called the optimum time step t_{step}, is obtained by rearranging Equation 5-27 to give:

$$t_{step} = \frac{z_{max} \ \Delta x \Delta y}{Q_{max}} \qquad (5\text{-}28)$$

where

t_{step} = time step over which sedimentation occurs
Q_{max} = maximum transport rate in entire grid area
z_{max} = maximum thickness of sediment eroded during one iteration established by user as input parameter DMOBILE

Thus, t_{step} is recalculated for each iteration and is based on the maximum transport rate found in the surf zone. Several important relationships arise from Equation 5-28:

(1) As transport rates Q increase, time step t_{step} decreases, thereby preventing unusually abrupt changes in submerged topography by limiting erosion that can occur during an iteration.

(2) Increasing the area of cells $\Delta x \Delta y$ allows more time to be represented by a single iteration.

(3) Increasing z_{max} by increasing the thickness of moving bedload DMOBILE provided as input allows longer spans of time to be represented by a single iteration.

Figure 5-14 shows WAVE's procedure for eroding sediment, where erosion is represented by decreases in thickness Δz of sediment in cells (Equation 5-27), accompanied by decreases in topographic elevation. The total change in thickness in a cell during one time increment is equal to the sum of changes in thickness of each of four grain types:

$$\Delta z = \Delta z_1 + \Delta_2 + \Delta z_3 + \Delta z_4 \qquad (5\text{-}29)$$

where

$$\Delta z \qquad = \quad \text{change in topographic elevation}$$
$$1, 2, 3, 4 \quad = \quad \text{four grain types}$$

Thus, volumes of grain types moving to and from cells (Equation 5-5) can be expressed as elevation changes Δz because cell areas $\Delta x \Delta y$ are constant, and changes in volumes are related to changes in topographic elevation by:

$$\Delta Vol = \Delta z\, \Delta x\, \Delta y \qquad (5\text{-}30)$$

As in Equation 5-9, topographic change caused by erosion is governed by transport efficiencies of each grain type, where decreases in thickness of a cell resulting from removal of any of four grain types is given by:

$$\Delta z_1 = \Delta z\, \varepsilon_1 \qquad (5\text{-}31)$$
$$\Delta z_2 = \Delta z\, \varepsilon_2$$
$$\Delta z_3 = \Delta z\, \varepsilon_3$$
$$\Delta z_4 = \Delta z\, \varepsilon_4$$

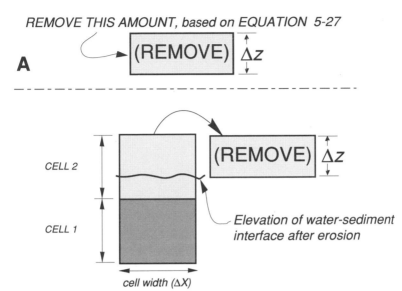

B Situation where total volume to be removed <u>is</u> available in upper cell.

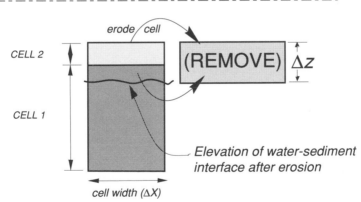

C Situation where total volume to be removed is <u>not</u> available in upper cell.

Figure 5-14 Schematic cross section of sediment cell illustrating WAVE's procedure for eroding sediment: (A) Volume of sediment to be eroded is expressed as thickness Δz obtained with Equation 5-27. (B) If thickness Δz is available in upper layer, up to four sediment types are eroded according to their individual transport efficiencies given by Equation 5-31. (C) If thickness in upper sediment cell is less than Δz entire cell is eroded and remainder of Δz is eroded from cell or cells beneath.

129

where

Δz = total thickness of eroded or deposited sediment, equal to the elevation change of sediment cell, where $+\Delta z$ denotes deposition and $-\Delta z$ denotes erosion

$\Delta z_{1,2,3,4}$ = change in thickness of each of four grain types in sediment cell

$\varepsilon_{1,2,3,4}$ = transport efficiencies of each of four grain types (Equation 5-23)

WAVE's procedures for simulating erosion begins by employing Equation 5-27 to determine the amount of sediment to be removed in a cell during one time increment, expressed as thickness Δz (Figure 5-14A). WAVE then adjusts thicknesses of cells (Figure 5-14B) to reflect erosion of the four grain types expressed as $\Delta z_1, \Delta z_2, \Delta z_3, \Delta z_4$, whose contribution to the total eroded thickness Δz are a function of their individual transport capacities (Equation 5-31). WAVE first erodes grain type 4, which has the smallest grain diameter, by amount Δz_4. Remaining grain types are treated similarly until thickness Δz has been removed. If grain type 4 is not available in sufficient amount to contribute the full amount Δz_4 as prescribed by Equation 5-31, thicknesses of other grain types are removed until thickness Δz has been removed. If amount Δz is not available in the uppermost cell (Figure 5-14C), cells underneath are eroded until Δz is removed or until all layers are eroded and "basement" is reached. "Basement" denotes substrate that cannot be eroded, whose topography is initially supplied as input. Changes in thicknesses of each of four grain types are stored in array REMOVE (Figure 5-15).

WAVE does not employ threshold velocities or shear stresses to determine whether sediment of a particular grain type is to be eroded. Instead WAVE employs Equations 5-10 through 5-23 to calculate transport efficiencies ε_1, ε_2, ε_3, and ε_4 of each of four grain types (Equation 5-23), which empirically govern the amount of sediment contributed by each during erosion. However, transport efficiencies are based on threshold shear stresses provided by Equation 5-13, so that a general relationship exists where grains with large threshold velocities have lower transport efficiencies than those with small threshold velocities and therefore contribute proportionally less to the volume eroded from a cell.

TRANSPORT AND DEPOSITION

WAVE represents sediment transport by moving volumes of sediment stored in array REMOVE to adjacent cells (Figure 5-15). WAVE expresses volumes of sediment stored in REMOVE as longshore QL and cross-shore QC components, employing directions of sediment transport α at each grid cell to calculate quantities of sediment transported represented by vectors having magnitude and direction:

$$QL_{ks_{i,j}} = \Delta Vol_{ks_{i,j}} (\cos \alpha_{i,j})^2 \tag{5-32}$$

$$QC_{ks_{i,j}} = \Delta Vol_{ks_{i,j}} (\sin \alpha_{i,j})^2 \qquad (5\text{-}33)$$

where

$QL_{ks_{i,j}}$ = component of moving grain type ks, parallel to y axis (Figure 5-15)
$QC_{ks_{i,j}}$ = component of moving grain type ks, parallel to x axis
$\Delta Vol_{ks_{i,j}}$ = total volume of moving grain type ks moving into cell i, j from four surrounding cells (along both x and y axes, Figure 5-15B)
ks = grain types $1, 2, 3$, and 4
i, j = cell indices
α = angular direction of sediment transport, based on directions of longshore currents (Figure 5-1A) provided by WAVECIRC (Figure 5-7).

If WAVE's circulation model is run for several hundred iterations to allow longshore currents to converge to stable numerical values (Figure 3-17), CURRENT then provides directions of sediment transport α in each cell from U and V components of longshore currents (Equations 3-64 and 3-65) that are provided in turn by WAVECIRC (Figure 5-1A and 5-7). Otherwise, WAVE can employ procedures in subroutine DIRECT to provide directions of sediment transport that are assumed to be parallel to submerged topography, so that calculation of longshore currents is not required.

Subroutine MOVE controls movement of longshore QL and cross-shore QC components of the total sediment volume Vol_{ij} moving to or from neighboring cells (Figure 5-15C). WAVE accounts for transport of four grain types by employing Equation 5-34 to represent volumes $Vol_{ks_{i,j}}$ that are stored in array REMOVE (Figure 5-15B) and are adjusted to reflect movement of sediment to or from adjacent cells indexed with the scheme in Figure 3-13:

$$Vol_{ks_{i,j}} = Vol_{ks_{i,j}} + QC_{ks_{i-1,j}} + QC_{ks_{i+1,j}} + QL_{ks_{i,j-1}} + QL_{ks_{i,j+1}} \qquad (5\text{-}34)$$

where

$Vol_{ks_{i,j}}$ = volume of moving grain type ks stored in array REMOVE at cell location i, j (Figure 5-15B)

Movement to and from cells can change the composition of sediment within cells, thus allowing WAVE to represent sorting and mixing by waves as they erode and transport sediment. Deposition is represented by adding sediment in array REMOVE to open sediment cells (Figure 5-4) of the four-dimensional array shown by Figures 5-15 and 5-3.

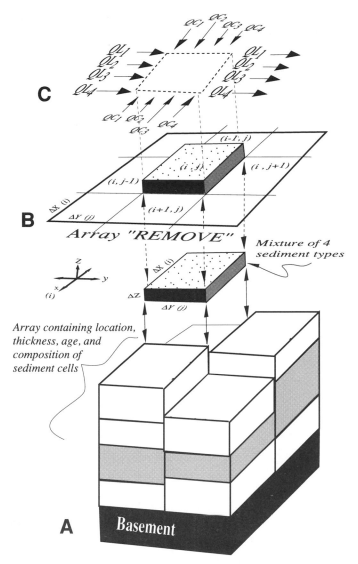

Figure 5-15 WAVE's procedure for eroding, transporting, and depositing sediment: (A) Volume of sediment in cell to be eroded can consist of up to four grain types contained in 4-D array representing aggregate volume of all sediment types previously deposited in cell. (B) Thicknesses of up to four grain types are eroded from cell are temporarily stored in array REMOVE, which represents temporary suspension of sediment in water column. (C) Transport is represented by movement of four grain types to and from cell, yielding revised proportions and thicknesses of grain types stored in array REMOVE. Revised thicknesses of grain types in REMOVE are then returned to 4-D array, thereby representing deposition. QL_1, QL_2, QL_3, and QL_4 denote volumes of four grain types moving in longshore or y-coordinate direction, whereas QC_1, QC_2, QC_3, and QC_4 denote volumes moving in cross-shore or x direction.

EXPERIMENTS DEMONSTRATING WAVE'S PROCEDURES FOR SIMULATING TRANSPORT

Two initial experiments demonstrate WAVE's procedures for simulating sediment transport. Experiment 1 involves a beach with constant slope, whereas Experiment 2 involves a beach having an irregular, undulating surface. The two experiments show the effects of contrasting topography on sediment transport, and their simplicity allows WAVE's procedures to be described more readily than with examples that attempt to parallel actual beaches.

Experiment 1: Beach with constant slope
Experiment 1 involves a beach with constant slope and an area 500 meters long and 250 meters wide. Topographic contours and direction of waves are shown in Figures 5-16 and 5-17, and input data are summarized in Table 5-2. Waves approached from the southeast with a wave angle of 150 degrees and wave height of 1.3 meters. At the outset, the eastern half of the beach was covered with a layer of medium sand two meters thick, whereas the western half was covered with a layer of fine sand two meters thick (Figure 5-17). Each layer consisted of four grain types in different proportions (Table 5-2), so that the average grain diameter was either "medium" or "fine" as defined by grain-size classifications in Table D-2 of Appendix D.

Plate 6 and Figure 5-17 present results after only a few iterations, and show that medium-grained sand (yellow) moved alongshore in the surf zone, replacing finer sand (blue) that moved parallel to shore. All movement was confined to the narrow surf zone. WAVE sorted the mixture of the moving bedload within the surf zone, as shown by color changes representing sediment composition. Plate 6 shows that the moving bedload on the west side of the beach became coarser (lighter blue), where sorting caused finer grains to move less than larger grains. By contrast, bedload on the east side became finer (lighter yellow) because the coarsest grains were not included in the moving bedload.

Plate 7 is an enlarged cross section of the surf zone in which changes in thickness and composition of the moving bedload are marked by color changes in the upper several centimeters of beach sediment. Plates 6 and 7 show that within the surf zone a thin layer of medium sand (yellow) moved from east to west and displaced finer sediment (blue).

Experiment 1 may be summarized as follows:

(1)   Shoaling wave heights and refraction angles were identical in each column of the grid (Figure 5-16), with columns defined in Figure 5-6.

(2)   The breaker zone occurred at the same depths in each column (Figure 5-16) because the refraction of waves was the same in each column.

133

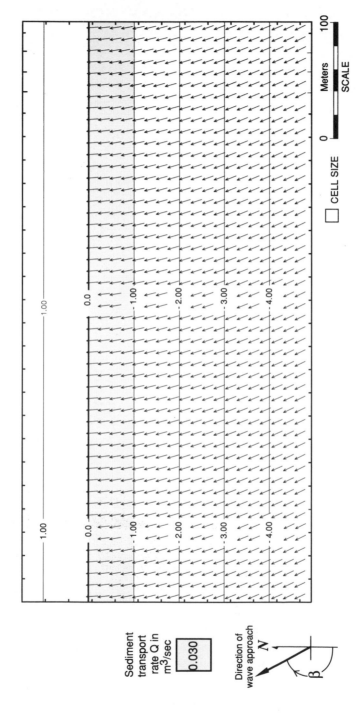

Figure 5-16 Map of Experiment 1 involving beach with constant slope. Contours show depths in meters. Wave orthogonals are shown with arrows. Sand transport rate was 0.03 cubic meters per second (shaded) in grid cells within surf zone. Transport rates are calculated on a cell-by-cell basis. Because beach is planar, waves broke at same location in each column of grid over width of surf zone. Sediment transport rates remained uniform along length of beach. Grid has 25 rows and 50 columns containing 1250 cells, each 10 meters square.

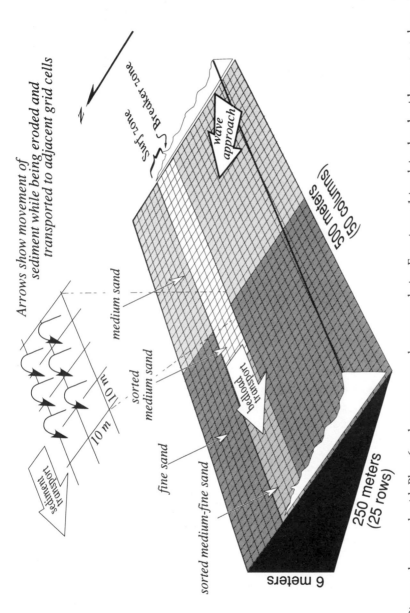

Figure 5-17 Diagram that accords with Plate 6 and represents sand transport during Experiment 1 involving beach with constant slope. Enlargement shows that bedload was confined to grid cells within surf zone and moved parallel to shore and topographic contours.

(3)   Wave energy at the breaker zone was also the same in each column because the refraction of waves was identical in each column.

(4)   The surf zone had the same width in each column (Figures 5-16 and 5-17, and Plate 6).

(5)   Longshore transport transport rates were identical within rows (Figure 5-16).

(6)   Longshore currents moved parallel to shore because the contours of submerged topography are parallel to shore (Figure 5-17).

(7)   The volume of sediment removed in each cell was transferred to its immediate downdrift neighbor in the same row (enlargement of Figure 5-17).

(8)   The volume of sediment removed in each cell was replaced by an identical volume of sediment from its immediate updrift neighbor (enlargement of Figure 5-17 and Plate 7).

(9)   The thickness of moving bedload was uniform within rows (Plate 7).

(10) The beach remained in equilibrium with unchanged topography, but composition of the bedload moving within surf zone changed because of sorting and mixing during erosion and transport (Plate 7).

Experiment 1 thus shows the response of a beach with constant slope, where conditions of refracting waves, width of surf zone, sediment transport rate, and sediment transport direction remained uniform in response to a uniform slope. While Experiment 1 demonstrates that WAVE's accounting procedures for eroding, sorting, transporting, and depositing sediment within the surf zone work effectively, its results are artificially simple because actual beaches rarely have constant slopes.

Experiment 2: Beach with irregular slope
Experiment 2 is more realistic because it demonstrates WAVE's reaction to an irregular slope. Input provided to Experiment 2 was identical to Experiment 1 (Table 5-2) except that an irregular slope and a wave angle of 200 degrees (Figure 5-18) were provided. Plates 8 through 11 show responses in Experiment 2, including patterns of refraction, sediment transport, and mixing that can be contrasted with Experiment 1. In Experiment 2, shoaling wave heights and refraction angles varied from cell to cell because submerged topography varied from cell to cell (Figure 5-18 and Plate 8). Waves broke at different water depths with different amounts of energy, causing transport rates to vary from cell to cell (Plate 9), thereby causing the thickness and composition of bedload to vary between cells (Plates 8, 10 and 11). Furthermore, in contrast to Experiment 1, the moving bedload in Experiment 2 moved

136

Table 5-2 Input data for Experiments 1 and 2, which are similar for both experiments except that Experiment 2 has deep water wave angle of 200 degrees and different beach topography.

Wave parameters:

Wave period (seconds)	13
Deepwater wave height (meters)	1.3
Wave angle (degrees)	150
	(200 degrees in Experiment 2)

Sediment transport:

Maximum thickness allowed for moving bedload (meters)	0.04
Coefficient *K* used in transport equation (Eq. 4-17)	0.77

Grid parameters:

Area of grid (meters)	250 x 500
Grid cell length (meters)	10
Number of rows *i*	25
Number of columns *j*	50

Sediment parameters: (left half of beach)

	Diameter (mm)	Density (kg/m3)	Percentage
Coarse sand	0.80	2650	10
Medium	0.40	2650	20
Fine	0.10	2650	30
Finest	0.05	2650	40

Sediment parameters: (right half of beach)

	Diameter (mm)	Density (kg/m3)	Percentage
Coarse sand	0.80	2650	20
Medium	0.40	2650	40
Fine	0.10	2650	20
Finest	0.05	2650	20

---

in various directions due to the irregular topography (enlargement of Figure 5-19 and Plate 9), so that the width of the surf zone did not remain uniform alongshore (Plates 9 and 10), and the moving bedload did not maintain a consistent average grain size as it moved alongshore (Plate 11). The moving bedload is generally shown by green in Plates 10 and 11, indicating that a mixture of mostly medium sand moved within the surf zone.

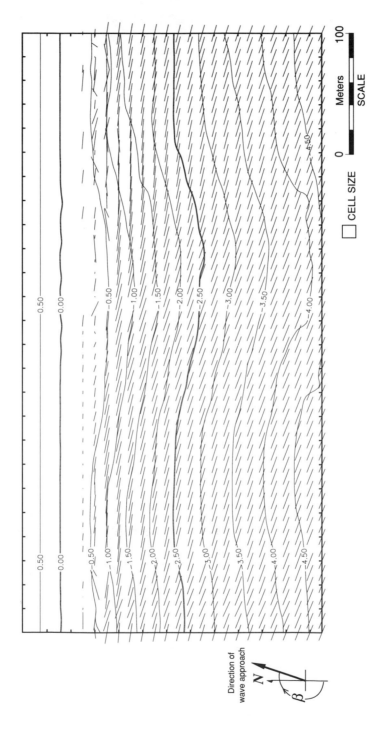

Figure 5-18 Experiment 2 involving beach with irregular slope. Wave crests are denoted by lines parallel to wave crests, with line lengths proportional to wave heights. Heights generally increase towards shore until waves break, and then decrease between breaker zone and shore. Grid has 25 rows and 50 columns containing 1250 cells each 10 meters square. Contours show depths in meters.

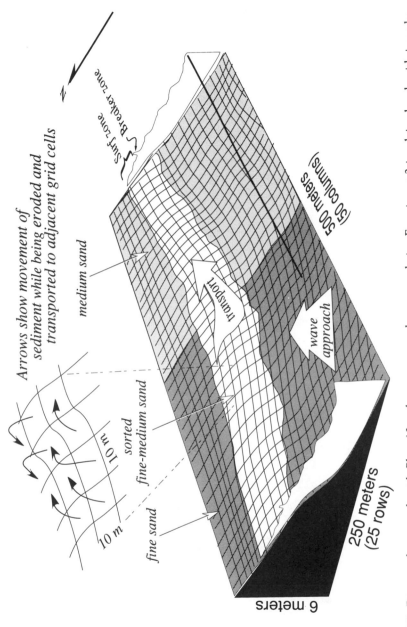

Figure 5-19 Diagram that accords with Plate 10 and represents sand transport during Experiment 2 involving beach with irregular slope. Enlargement shows that bedload moved in various directions in response to irregular slopes.

Figure 5-20 helps to explain why the composition and thickness of moving bedload differs between Experiments 1 and 2. In Experiment 1, sediment moving from a cell was always replaced by an identical amount of sediment moving from adjacent cells, thereby maintaining uniform composition and thickness (Figure 5-20A) through-

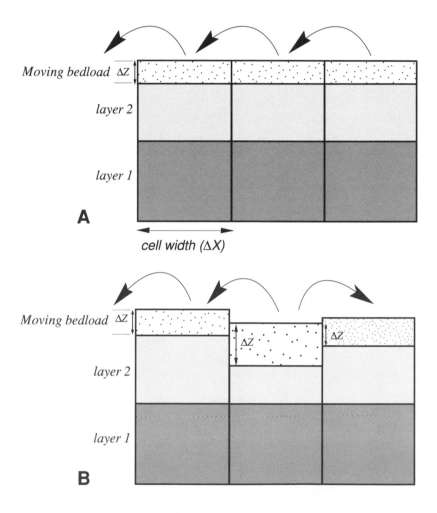

Figure 5-20 Schematic cross sections through sediment cells contrasting thickness of bedload moving from cell to cell in surf zone in Experiments 1 and 2: (A) In ideal planar beach of Experiment 1, thickness of moving bedload Δz did not change because sand transport rates were uniform within surf zone (Figure 5-16), signifying that sediment eroded from specific cell during erosion cycle was immediately replaced by equal amount that entered from adjacent cell during deposition cycle. (B) In Experiment 2 sand transport rates varied over irregular slope (Plate 9), causing thickness of moving bedload to vary from cell to cell (Plate 11).

out the experiment. In Experiment 2, waves broke with variable intensity as topography changed from cell to cell, causing transport rates to vary from cell to cell, thereby causing the thickness of moving bedload to vary from cell to cell (Figure 5-20B). Thus, while Experiments 1 provides a simple demonstration of WAVE's procedures for representing sediment transport, Experiment 2 shows that WAVE can realistically represent longshore transport on beaches whose irregular topography allows for a more dynamic interaction between wave processes, beach topography, and sediment transport.

SUMMARY

WAVE's procedures for erosion, transport, and deposition operate in a cellular grid where sediment cells represent layers composed of a mixture of four grain types. Sorting is represented by moving grain types as functions of their diameters and densities. Parameters including Shields parameter, Reynolds number, pivot angle, and fall velocity are used to calculate transport efficiencies of each of four grain types, which in turn determine the relative contributions of each grain type in the moving bedload. Erosion is represented by removing each of up to four grain types from sediment cells, whereas transport is represented by moving grain types to or from adjacent cells, and deposition is represented by returning grain types to sediment cells. Accounting procedures provide a three-dimensional record of sedimentary deposits providing a stratigraphic history of depositional and erosional events that can be displayed later. While Experiments 1 and 2 are simple examples of WAVE's representation of wave-induced erosion, transport, and deposition, subsequent experiments show that WAVE is effective in simulating nearshore transport by waves in realistic detail as input parameters are provided to more accurately represent conditions of actual beaches, and simulations span longer periods of time.

141

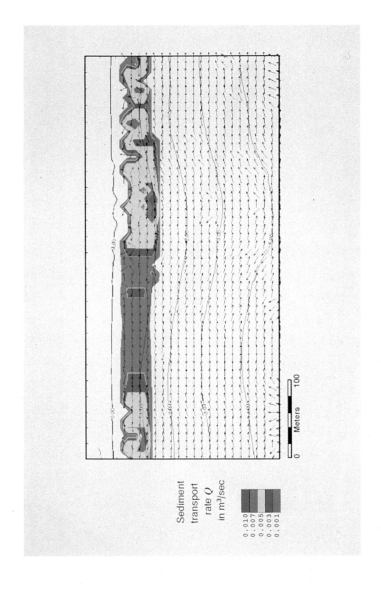

Sediment
transport
rate Q
in m³/sec

0.010
0.007
0.005
0.003
0.001

0   Meters   100

Plate 9 Experiment 2. Beach with irregular slopes. Map shows depths with contours in meters, directions of longshore transport with arrows, and rate of longshore transport with colors. Transport rates in m³/sec are shown on cell-by-cell basis.

50 m

50 m

5.0 m

x10

Plate 10 Experiment 2. Beach with irregular slopes. Perspective display shows composition of sand at surface of beach after 24 hours, according with conditions represented in Figure 5-19. Very-fine sand (blue) covers west half of beach, whereas medium sand (yellow) covers east half. Location and width of surf zone are denoted by discoloration caused by sorting and mixing of sediment by longshore transport. Bedload consists mostly of fine sand (green), that moved from left to right within surf zone. Waves approached from lower-left corner. Vertical scale is 10 times horizontal. Dimensions of beach area are 250 m by 500 m. Red box denotes area shown in Plate 11.

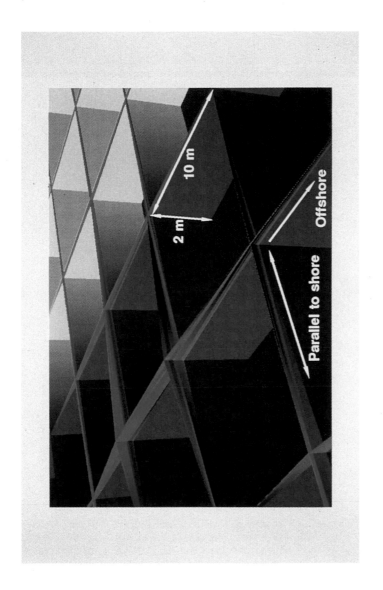

Plate 11 Experiment 2- Beach with irregular slopes. Enlarged perspective fence display showing composition of sediment layers deposited by waves. Location of enlarged area is shown by red box in Plate 10. Colors represent a range of grain sizes between medium sand (yellow) through very-fine sand (blue). Fine and very-fine sand (green and blue) preferentially moved as bedload above underlying sediment. Thickness of upper layers varies from cell to cell as a result of variations in topography and of transport rates that vary from cell to cell. Distance between sections is ten meters, and thickness of sand fill at outset is two meters. Vertical scale is 2.5 times horizontal.

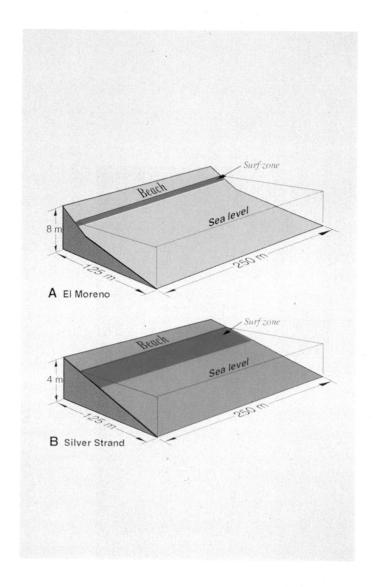

Plate 12 Experiments 3 and 7- El Moreno and Silver Strand Beaches. Perspective views comparing slopes, surf zone width, and composition of sand in surf zone after one week of simulated time. Yellow, green, and blue represent end members consisting of coarse, medium, and fine sand, respectively. Blue bands show location of surf zone, where longshore transport has sorted and lifted finer sands into moving bedload, leaving coarser sands buried. (A) El Moreno Beach has stepped topography and narrow surf zone indicated by narrow band of blue. (B) Silver Strand Beach has gentle slope and wider surf zone. Figures are redrafted from computer-generated displays.

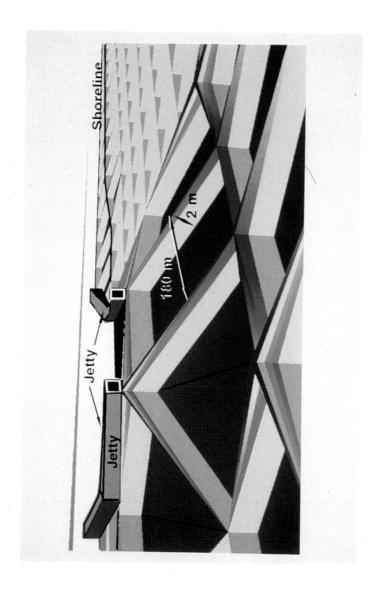

Plate 13 Experiment 8. Anaheim Bay jetty. Enlarged perspective fence display showing composition of sediment layers deposited by waves around jetty. Colors denote a range of grain sizes between medium sand (red) and fine sand (blue). Note coarsening-upward trend of sediment accumulating against jetty. Location of jetty (grey barriers) and shoreline (black line) are among drafted additions to display. Sections are 180 meters apart. Vertical scale is 30 times horizontal.

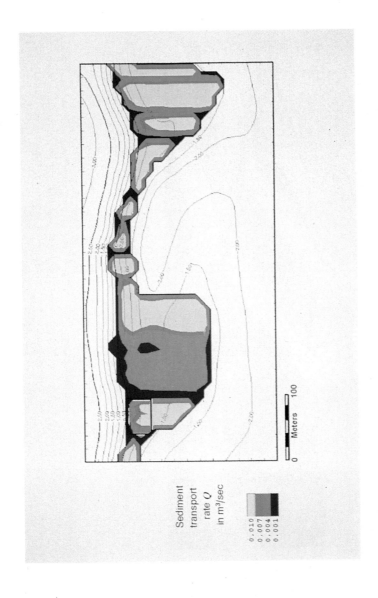

Plate 14 Experiment 9. Beach with longshore bar, southern Oregon. Map shows rates of sand transport with colors and topography with contours in meters. Colors show areas in surf zone where longshore transport rates range between 0.001 and 0.01 m^3/sec. Rates are greatest and surf zone widest over crests of longshore bars. Grid representing area has 30 rows and 60 columns containing 1800 grid cells, each 10 meters square. Red box denotes area shown in Plate 15.

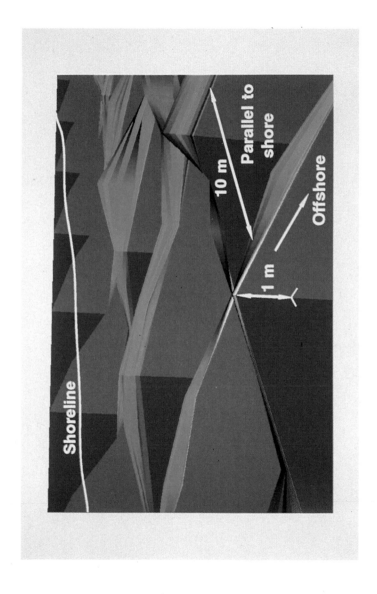

Plate 15 Experiment 9. Beach with longshore bar, southern Oregon. Enlarged perspective fence display showing composition of deposits on bar crest, at location shown by red box in Plate 14. Layers of sediment are shown by thin bands of color denoting a range of grain sizes between coarse sand (yellow) and fine sand (blue). Coarsest deposits (yellow, and green) occur on bar crest. Vertical scale is 3 times horizontal.

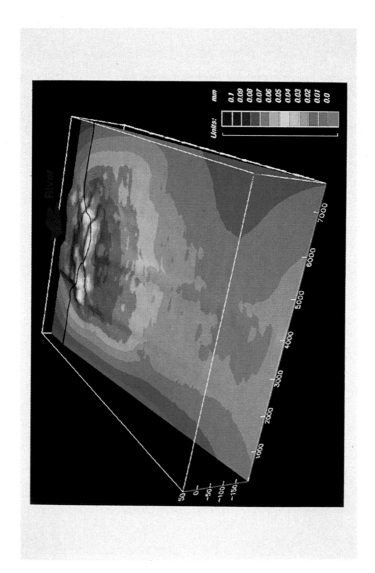

Plate 16 Experiment 10- Delta without waves. Perspective view in which colors denote grain size at surface. Colors represent a range of grain sizes between fine sand (red) through fine silt (blue). Arrow shows path of river and black line shows position of shoreline. Vertical scale is eight times horizontal. Overall dimensions of block are 7.5 km by 10 km by 250 m.

CHAPTER 6

Simulating
Longshore Transport
on Beaches

This chapter describes experimental simulations of longshore transport along beaches and presents experiments that are compared with actual field studies. Here, WAVE has been operated as a stand-alone program without fluvial deposition by SEDSIM. The results show that WAVE can operate at scales ranging from small areas and short time spans, to large areas and long time spans. Experiments include discussion of input parameters, with results shown with maps and three-dimensional diagrams that display topography, wave orthogonals, sediment transport rates, and sediment composition.

EL MORENO AND SILVER STRAND BEACHES

Measurements of wave properties and nearshore sand transport rates at El Moreno and Silver Strand Beaches in southern California (Figure 4-3) by Komar and Inman (1970) provide "classic" data for testing and calibrating simulation experiments involving WAVE. At El Moreno and Silver Strand Beaches, Komar and Inman measured transport rates by placing dyed sand in trenches during low tide and observing its dispersion during high tide (Figure 4-4). From their observations of local wave heights, breaking-wave angles, longshore currents, and they formulated the wave-power equation (Equation 4-17) that relates longshore transport rates to wave energy. Because the wave-power equation is incorporated in WAVE, data from El Moreno and Silver Strand beaches are particularly suitable for comparisons with experiments involving WAVE.

Experiment 3: Longshore transport at El Moreno Beach

Experiment 3 involving El Moreno Beach employed data shown in Table 6-1, whose entries were averaged from observations by Komar and Inman provided in Table 4-4, and Tables A-5, A-6, and A-7 of Appendix A. Komar and Inman observed that waves at El Moreno Beach were generated from sea breezes that had short periods, small heights, and an average incidence angle of ten degrees. The beach had a steep shoreface about 30 meters wide that extended into a broad shallow tidal terrace. Similarly, Experiment 3 involved a steplike topography (Figures 6-1 and Plate 12A) constructed

Table 6-1 Experiment 3: Input data for simulating El Moreno Beach.

Wave properties:

Wave period (seconds)	3
Deepwater wave height (meters)	0.3
Deepwater wave angle (degrees)	200

Sediment transport:

Maximum thickness of moving bedload (meters)	0.084
Coefficient K used in transport equation (Equation 4-17)	0.77

Grid parameters:

Area of grid (meters)	125 x 250
Grid cell length (meters)	5
Number of rows i	25
Number of columns j	50

Sediment parameters

Layer 1 (lower layer) at outset of experiment:
Thickness = 1.0 meter, average grain diameter = 0.60 mm:

	Diameter (mm)	Density (kg/m^3)	Percentage
Coarsest	0.65	2650	40
Medium	0.60	2650	40
Fine	0.55	2650	10
Finest	0.40	2650	10

Layer 2 (upper layer) at outset of experiment:
Thickness = 1.0 meter, average grain diameter = 0.60 mm:

	Diameter (mm)	Density (kg/m^3)	Percentage
Coarsest	0.65	2650	40
Medium	0.60	2650	40
Fine	0.55	2650	10
Finest	0.40	2650	10

144

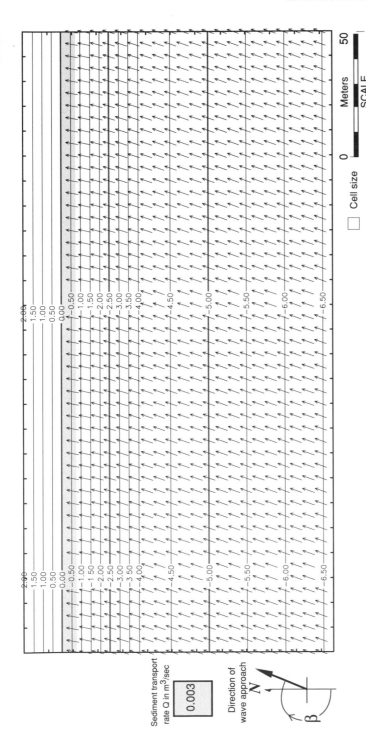

Figure 6-1 Experiment 3 - El Moreno Beach: Map shows refraction of wave orthogonals (arrows) as waves moved toward shore. Sediment transport rates within surf zone (shaded) were approximately 0.003 m³/sec for deep-water wave height of 0.3 meters. Surf zone was five meters (one cell) wide. Deep-water wave angle β was 200 degrees, using convention shown in Figure 3-13. Contours in meters show topographic elevations with respect to mean sea level. Shading represents sediment transport rate, which is specifically 0.003 m³/sec per cell area, with cell size shown by labelled box.

from a cross section provided by Komar (1969). The simulation included an area of 125 by 250 meters represented by a grid divided into 25 rows and 50 columns, incorporating 1250 cells, each 5 meters square. At the outset the area was covered with mixed coarse quartz-feldspar sand having an average grain size of 0.6 millimeters.

Experiment 3 involved 10,000 iterations representing about three months of simulated time. Results are shown in Figure 6-1, where shoaling, refracting waves are represented by arrows that represent wave orthogonals. Simulated wave heights, angles, and transport rates are presented in Table 6-2 for comparison to those observed by Komar and Inman (1970). The simulated longshore transport rate of 285 cubic meters per day is within ranges reported Komar and Inman and compares favorably with their average value of 225 cubic meters per day. The width of the simulated surf zone, however, was narrower than those observed by Komar and Inman because simulated waves had identical wave heights, angles, and periods throughout the experiment, causing them to break at the same depths. If variable wave heights, angles, and periods were provided as input, the simulated waves would break over a range of water depths, producing a less-distinct breaker zone and wider surf zone more similar to that actually observed at El Moreno Beach.

Table 6-2 Comparison of parameters observed at El Moreno Beach with those obtained in Experiment 3. Observed values are averages from Komar and Inman (1970) provided in Tables 4-4, A-5, and A-7.

	Observed	Simulated by WAVE
Breaking wave heights (meters)	0.3	0.3
Breaking wave angle (degrees)	9	9.5
Width of surf zone (meters)	17	5
Longshore transport rate (m^3/day)	225	285
Longshore transport rate (m^3/sec)	0.0026	0.0033

Because Experiment 3 involved a simple sloping beach, its longshore transport properties were similar to Experiment 1 (Figure 5-17 and Plates 6 and 7), with the simulated bedload moving parallel to shore and maintaining a constant thickness throughout the experiment. Similarly, the upper few centimeters of sediment changed composition as a result of mixing and sorting caused by longshore transport in the surf zone, as shown by Plate 12A, where color variations indicate differences in composition.

Experiment 3 documents that WAVE can simulate longshore transport rates similar to those observed at El Moreno Beach. However, WAVE can also simulate transport rates for wave conditions not observed by Komar and Inman and could be useful for analyzing the sensitivity of longshore transport rates to wave conditions for which field data are unavailable. Experiments 4, 5, and 6 that follow provide examples where WAVE is used to forecast transport rates under conditions of varying wave heights, angles, and periods, thereby demonstrating the flexibility and utility that computer models can provide as predictive tools.

Experiment 4: Increasing wave heights at El Moreno Beach

In Experiment 4, deep water wave height H_O was increased from 0.3 to 0.9 meters in seven steps (Figure 6-2), with other input data from Experiment 3 (Table 6-1) unchanged. A plot of transport rates versus wave height in Figure 6-2 shows that a threefold increase in wave heights caused transport rates to increase fourfold. Similarly, as wave heights were increased, the width of the surf zone increased as larger waves broke in deeper water. Figure 6-3 shows the distribution of longshore transport rates and the surf-zone width obtained for the case when wave height was 0.7 meters. Comparison of Figures 6-1 and 6-3 shows that increasing wave height from 0.3 meters (Experiment 3) to 0.7 meters (Experiment 4) caused transport rates to exceed 0.01 cubic meters per second per cell and resulted in a fourfold increase in surf-zone width. Thus, Experiment 4 indicates that increases in wave heights can cause significant increases in transport rates and can cause the surf zone to widen as waves break in deeper water.

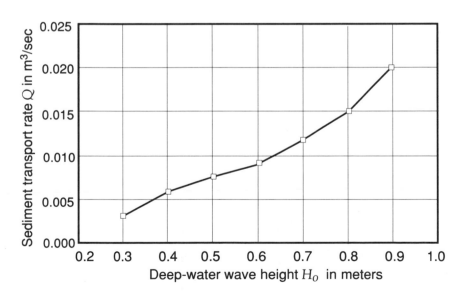

Figure 6-2 Experiment 4 - El Moreno Beach: Graph of deep-water wave height H_O versus sediment transport rate Q for seven wave heights provided as input, with other input parameters (Table 6-1) unchanged.

147

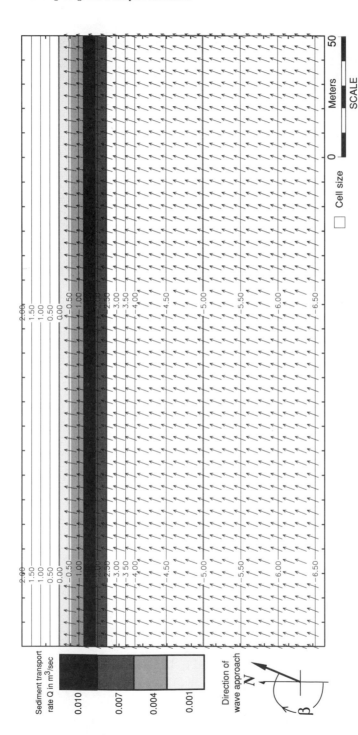

Figure 6-3 Experiment 4 - El Moreno Beach: Increased wave energy resulting from increased wave height of 0.7 meters produced wider surf zone and greater sediment transport rates as compared with Figure 6-1, where wave height was 0.3 meters. Shaded areas show that transport rates ranged between 0.001 and 0.010 m³/sec within surf zone, which was 20 meters (four cells) wide. Arrows represent wave orthogonals, and contours show topographic elevation in meters with respect to mean sea level. Deep-water wave angle was 200 degrees. Four levels of shading pertain to four specific transport rates shown. Transport rates are assigned on a cell-by-cell basis and do not form a continuum. Shaded bands, therefore, do not constitute contour intervals.

Experiment 5: Changing wave angle at El Moreno Beach

In Experiment 5, deep-water wave angle β was increased from 180 to 250 degrees in ten steps (Figure 6-4), with other input data from Experiment 3 (Table 6-1) unchanged. The plot of transport rates versus wave angle in Figure 6-4 shows that changes in wave angle affected transport rates less than changes in wave height described in Experiment 4. Wave angles between 180 and 250 degrees caused transport rates to range between 0.0 and 0.0047 cubic meters per second, in contrast to large increases in transport rates caused by increasing wave heights. Transport rates were highest when the deep-water waves approached at 220 degrees, where after refracting five degrees, they arrived at the breaker zone with an incidence angle of 45 degrees. Similarly, Longuet-Higgins' (1970) provides theoretical arguments suggesting that transport rates are largest for waves breaking with incidence angles of 45 degrees.

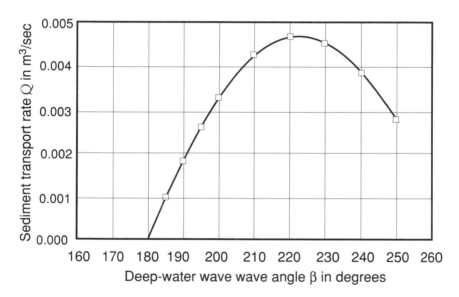

Figure 6-4 Experiment 5 - El Moreno Beach: Graph of deep-water wave angle β versus sediment transport rate Q for ten wave angles provided as input, with other input parameters (Table 6-1) unchanged. Maximum rate occured when deep-water wave angle was 220 degrees because deep-water waves refracted by five degrees before reaching breaker zone, so that incidence angle at breaker zone relative to shore was at an optimal (Longuet-Higgins, 1970) 45 degrees.

Experiment 6: Increasing wave periods at El Moreno Beach

In Experiment 6, wave period T was increased from three to ten seconds in seven steps (Figure 6-5), with other input data (Table 6-1) unchanged. Figure 6-5 shows that transport rates increased sharply when wave periods increased from three to six seconds, but changed little thereafter. Relationships between longshore transport and

149

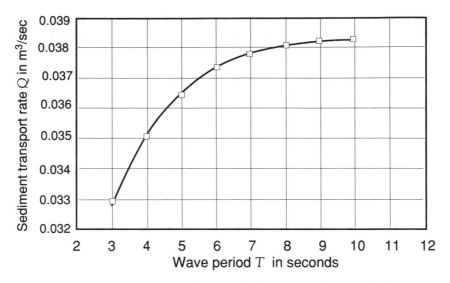

Figure 6-5 Experiment 6 - El Moreno Beach: Graph of wave height T versus sediment transport rate Q for eight wave periods provided as input, with other input parameters (Table 6-1) unchanged.

wave period are nonlinear, and the experiment documents that computer procedures like those employed in WAVE can forecast changes that may not be intuitive.

Results of Experiments 3, 4, 5, and 6 suggest that WAVE could be applied to coastal engineering problems, where simulations could forecast rates of sand transport caused by changes in wave properties. At El Moreno Beach for example, WAVE predicted that changes in wave height would cause the greatest change in sand transport rates, whereas changes in wave period would have only minor effects. However, other combinations of wave heights, angles, and periods could be used to forecast longshore transport rates under conditions for which observational data are lacking, further emphasizing the role that computer procedures can play in the study of nearshore processes.

Experiment 7: Silver Strand Beach

Komar and Inman (1970) compared the wave climate at Silver Strand Beach (Figure 4-3) to that at El Moreno Beach and found that while their wave climates were much different, transport rates were similar. Waves at Silver Strand Beach had periods and heights (Table 4-5) that were three to four times greater than at El Moreno Beach (Table 4-4), but had low angles of incidence (three to five degrees) over a low slope (0.034), which dampened much of the longshore component of wave energy. By contrast, waves at El Moreno were smaller and approached at higher angles (nine to ten degrees) over a steeper slope (0.138), which allowed most of the wave energy to be directed alongshore. The interdependence between wave energy, slope, and incidence angle produced similar transport rates even though the two beaches had different wave properties. Experiment 7 provides a similar comparison with Experiment 3, where simulations of Silver Strand and El Moreno Beaches yielded similar results, with each beach having similar longshore transport rates despite differences in wave conditions.

Experiment 7 involving Silver Strand Beach employed data shown in Table 6-3, whose entries were averaged from Komar and Inman's (1970) observations provided in Tables 4-5, A-5, A-6, and A-7. Experiment 7 involved a gently dipping shoreface

Table 6-3   Experiment 7: Input data for simulating Silver Strand Beach.

Wave parameters:

Wave period (seconds)	11
Deepwater wave height (meters)	0.7
Deepwater wave angle (degrees)	190

Sediment transport:

Maximum thickness of moving bedload (meters)	0.046
Coefficient K used in transport equation (Equation 4-17)	0.77

Grid parameters:

Area of grid (meters)	125 x 250
Grid cell length (meters)	5
Number of rows i	25
Number of columns j	50

Sediment parameters:
Layer 1 (lower layer) at outset of experiment:
Thickness = 1.0 meter, average grain diameter = 0.175 mm:

	Diameter (mm)	Density (kg/m^3)	Percentage
Coarsest	0.20	2650	40
Medium	0.18	2650	40
Fine	0.15	2650	10
Finest	0.07	2650	10

Layer 2 (upper layer) at outset of experiment:
Thickness = 1.0 meter, average grain diameter = 0.175 mm:

	Diameter (mm)	Density (kg/m^3)	Percentage
Coarsest	0.20	2650	40
Medium	0.18	2650	40
Fine	0.15	2650	10
Finest	0.07	2650	10

151

whose topography is shown in Figures 6-6 and Plate 12B. The simulation included an area of 125 by 250 meters represented by a grid divided into 25 rows and 50 columns, incorporating 1250 cells, each 5 meters square. At the outset, the area was covered with a mixture of fine quartz-feldspar sand with an average grain size of 0.175 millimeters.

Results are shown in Figure 6-6, with wave orthogonals represented by arrows and submerged topography by contours. Shaded areas in Figure 6-6 denote the location of the surf zone, and the degree of shading denotes the magnitude of transport rates. Because the surf zone was five cells wide, the transport rate in each cell contributes to the total transport rate along the width of the surf zone. Figure 6-6 shows that transport rates within the surf zone were distributed among cells as a function of bottom shear stress, yielding a skewed normal distribution of transport rates similar to that on actual beaches (Figures 4-12 and 4-13A). Table 6-4 compares field measurements of Silver Strand Beach with simulation results of Experiment 7, documenting that the simulated wave properties and transport rates are comparable to those observed at Silver Strand Beach.

Table 6-4 Comparison of parameters observed at Silver Strand Beach with those obtained in Experiment 7. Observed values are averages from Tables 4-5, A-5, and A-7.

	Observed	Simulated by WAVE
Breaking wave heights (m):	0.7	0.7
Breaking wave angle (degrees):	5	5.5
Width of surf zone (meters)	75	25
Longshore transport rate (m³/day)	280*	432 **
Longshore transport rate (m³/sec)	0.0032*	0.005 **

*   excludes one extreme value of 0.03 m³/sec (Table A-7), otherwise average of observed rates would be 0.01 m³/sec, or 864 m³/day.

**   transport rate is distributed among 5 cells within surf zone (Figure 6-6), for each column of grid.

After a few hours of simulated time, the composition of the upper few centimeters of sand in Experiment 7 changed as a result of mixing and sorting by longshore transport the in surf zone, as shown by Plate 12B, where color differences denote differences in sediment composition. The surf zone in Experiment 7 representing Silver Strand Beach (Figure 6-6 and Plate 12B) was about five times wider than the surf zone in Experiment 3 that represented El Moreno Beach (Figure 6-1 and Plate 12A), generally according with Komar and Inman's observations. In Experiment 3, the surf zone was narrow because waves were small and the beach was steep, causing waves to break close to shore. In Experiment 7 the surf zone was wider because waves were larger and broke farther offshore in deeper water.

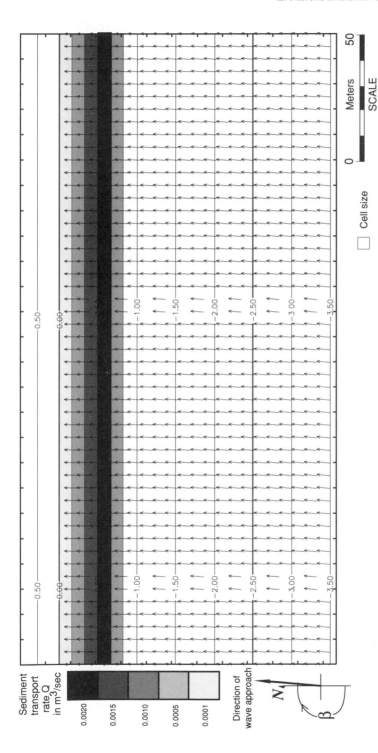

Figure 6-6 Experiment 7 - Silver Strand Beach: Transport rates (shaded) in surf zone varied between 0.0001 and 0.002 m³/sec for deep-water wave angle of 190 degrees and deep-water wave height of 0.65 meters. Surf zone was 25 meters (five cells) wide. Contours in meters show topographic elevation relative to mean sea level. Four levels of shading pertain to four specific transport rates shown. Transport rates are assigned on a cell-by-cell basis and do not form a continuum. Shaded bands, therefore, do not constitute contour intervals.

Figure 6-7 compares observed and simulated transport rates at El Moreno and Silver Strand Beaches. Simulated rates are within ranges observed at both beaches, and are similar even though their wave climates differ, in accord with observations of Komar and Inman. Figure 6-8 provides a more inclusive summary of transport rates measured at El Moreno and Silver Strand Beaches, as well as beaches in Florida and California observed by Watts (1953), Caldwell (1956), and Ingle (1966), further demonstrating that simulated rates in Experiments 3 and 7 are similar to those at actual beaches, most of which have rates that are generally less than 0.005 cubic meters per second, or 432 cubic meters per day.

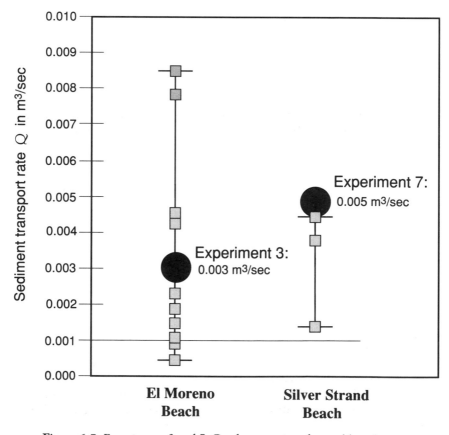

Figure 6-7 Experiments 3 and 7: Graph comparing observed longshore transport rates (squares) at El Moreno and Silver Strand Beaches (Table A-7) with simulated rates (solid circles).

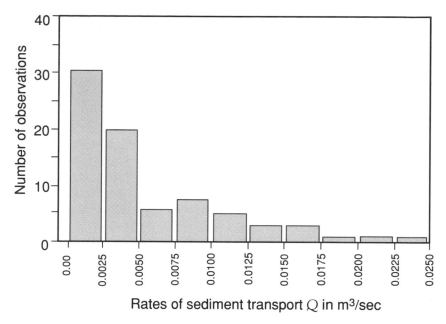

Figure 6-8 Histogram of aggregate of 80 measurements of longshore transport rates reported by Watts (1953), Caldwell (1956), Ingle (1966), and Komar (1969) and reproduced in Appendix A.

SIMULATING LONGSHORE TRANSPORT NEAR JETTIES AT ANAHEIM BAY

Jetties affect wave climate and longshore transport by interrupting littoral drift and causing sediment to accumulate updrift, while starving the coastline downdrift (Figure 6-9). Thus, jetties can change beach topography, and their construction can cause unanticipated and undesirable changes to the surrounding shore. Experiment 8 demonstrates that computer procedures like those incorporated in WAVE are potentially useful for forecasting the effects of jetties in altering longshore transport.

Experiment 8 simulated sand transport around jetties at Anaheim Bay in southern California (Figure 1-2 and Plate 1), where Caldwell (1956) estimated longshore transport rates by measuring changes in depths along transects oriented perpendicular to shore (Figure 4-3B). Caldwell provides averages of transport rates for intervals ranging from two to three months that were spaced over a span of one year (Tables 4-3, A-3, and A-4). Table 6-5 summarizes input data used in the experiment, which represents an area of 2700 by 9000 meters on a grid with 15 rows and 50 columns containing 750 cells, each 180 meters square. At the outset, the simulated area was covered with a mixture of medium-grained quartz-feldspar sand with average grain size of 0.35 millimeters. Submerged topography west of the jetty shown in Figure 6-10 was

155

Table 6-5 Experiment 8: Input data for simulating Anaheim Bay jetties.

Wave parameters:

Wave period (seconds)	15
Deepwater wave height (meters)	0.4
Deepwater wave angle (degrees)	220

Sediment transport:

Maximum thickness of moving bedload (meters)	0.08
Coefficient K used in transport equation (Equation 4-17)	0.77

Grid parameters:

Area of grid (meters)	2700 x 9000
Grid cell length (meters)	180
Number of rows i	15
Number of columns j	50

Sediment parameters:

Layer 1 (lower layer) at outset of experiment:
Thickness = 1.0 meter, average grain diameter = 0.35 mm:

	Diameter (mm)	Density (kg/m^3)	Percentage
Coarsest	0.45	2650	40
Medium	0.30	2650	40
Fine	0.25	2650	10
Finest	0.15	2650	10

Layer 2 (upper layer) at outset of experiment:
Thickness = 1.0 meter, average grain diameter = 0.35 mm:

	Diameter (mm)	Density (kg/m^3)	Percentage
Coarsest	0.45	2650	40
Medium	0.30	2650	40
Fine	0.25	2650	10
Finest	0.15	2650	10

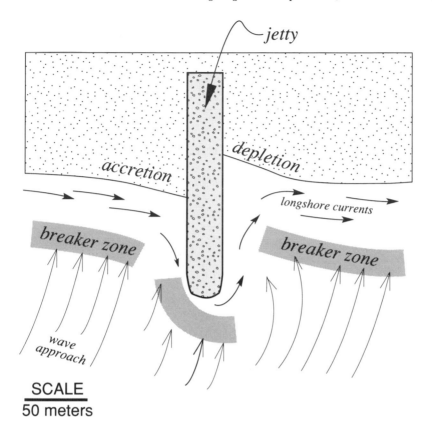

Figure 6-9 Map of hypothetical beach and jetty showing accretion of sand updrift, and depletion downdrift from jetty. Modified from Ingle (1966).

extrapolated from Caldwell's map (Figure 4-3B), allowing the simulated jetty to be near the center of the grid to avoid boundary effects.

Figures 6-10 and 6-11 show wave orthogonals, transport rates, surf-zone width, and transport directions after ten simulated months. Rates shown in Figure 6-10 are most variable along the eastern downdrift edge of the jetties, where wave crests exhibit their highest degree of bending or refraction (denoted by arrows). Simulated refraction patterns of Figure 6-10 are similar to those observed around actual jetties, as shown schematically in Figure 6-9.

Table 6-6 compares Caldwell's field observations with those simulated by WAVE. The simulated transport rates varied along the beach, but conformed closely to Caldwell's observations, as shown by the plot in Figure 6-12.

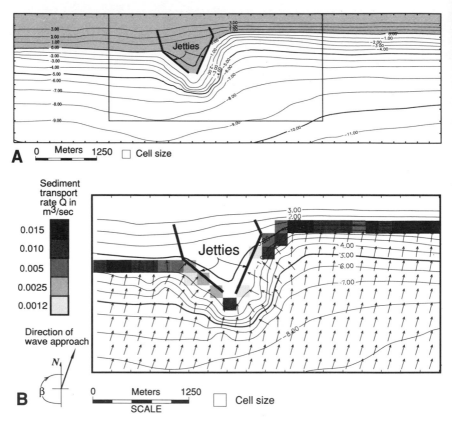

Figure 6-10 Experiment 8 - Anaheim Bay jetties. Contours in meters show topographic elevation with respect to mean sea level: (A) Map of simulated area at outset of experiment. Enlarged area in B shown by outline. (B) Topography, transport rates, and wave orthogonals (arrows) after one year of simulated time. Transport rates (shaded) in surf zone ranged between 0.0012 and 0.0150 m³/sec for deep-water wave angle of 220 degrees and deep water wave height of 0.4 meters.

Table 6-6 Comparison of parameters observed at Anaheim Bay jetties with those obtained during Experiment 8. Observed values are averages from Tables 4-3 and A-4. Simulated values are averaged along length of simulated beach (Figure 6-10B).

	Observed	Simulated by WAVE
Breaking wave heights (m):	0.5	0.3
Breaking wave angle (degrees):	5	9
Width of surf zone (meters)	75	less than 180 m.
Longshore transport rate (m³/day)	854	864
Longshore transport rate (m³/sec)	0.0099	0.010

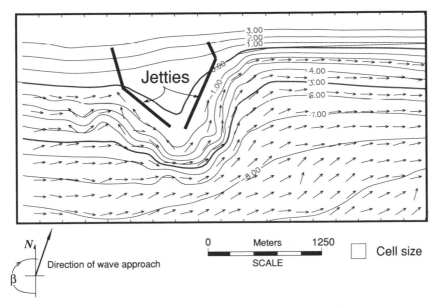

Figure 6-11 Experiment 8 - Anaheim Bay jetties: Directions of longshore transport (arrows) after one simulated year. Contours in meters show topographic elevation relative to mean sea level. Deep-water wave angle was 220 degrees.

After ten simulated months, composition of the upper few centimeters of sand changed in the surf zone because of mixing and sorting by longshore transport, as shown by Plate 2. Sand was depleted on the eastern side of the simulated jetties (Figures 6-10 and 6-11), similar to that Anaheim Bay (Figure 4-3B), where the Corps of Engineers placed 918,000 cubic meters of sand to compensate for depletion. Plate 13 is one a series of fence displays obtained in the experiment that can be viewed in rapid succession on a graphics workstation to show changes sediment accumulation around the jetties.

Longshore transport rates in Experiment 8 could be better simulated with smaller cells. Each cell in Experiment 8 was 180 by 180 meters (Table 6-5), whereas the surf zone was only between 50 and 100 meters wide, considerably narrower than a single cell 180 meters wide. Thus, estimates of rates would be improved if cells were smaller than the width of the surf zone. If cells were 50 meters square or less, the depth of breaking waves and width of the surf zone would be represented more accurately. However, Equation 6-1 shows that decreasing cell size $\Delta y \, \Delta x$ increases computing time because the amount of time represented during each time step t_{step} decreases with decreasing cell size:

$$t_{step} = \frac{z_{max} \bullet \Delta x \Delta y}{Q_{max}} \qquad (6\text{-}1)$$

where

t_{step} = time step over which sedimentation occurs
Q_{max} = largest volumetric transport rate in surf zone
Δz = change in surface elevation, or thickness, of sediment cell (meters)
$\Delta y, \Delta x$ = cell dimensions along x and y axes (meters)
z_{max} = maximum thickness of moving bedload, provided as input

Thus, while smaller cells increase accuracy, the time represented by each solution (iteration) of Equation 6-1 decreases, requiring more iterations to represent a given span of time. Fortunately, in simulating small areas we are generally concerned with short-term changes, and experiments with fewer iterations are suitable. Larger-scale simulations generally involve longer simulated time, but may require less detail and can employ larger cells that allow longer time steps and an acceptable number of iterations.

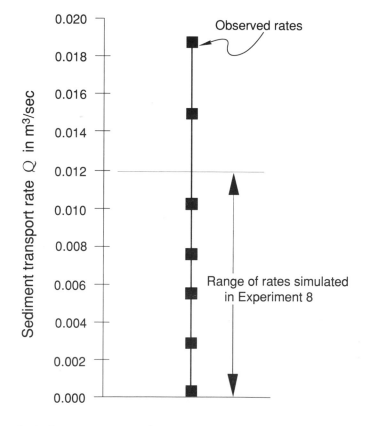

Figure 6-12 Experiment 8 - Anaheim Bay jetties: Graph comparing ranges of simulated transport rates with longshore transport rates observed by Caldwell (1956) and presented in Table A-4.

Of course, compromise is required to balance computer time with resolution requirements.

Experiment 8 posed difficulties for WAVE because it involved complex beach topography, unusually shaped jetties, and a large area. Here, WAVE's finite-difference schemes (Equation 3-75) underwent a difficult test because cells that represent the jetty were elevated above sea level and WAVE's finite-difference procedures can involve only neighboring cells that are submerged, while disregarding those above sea level. Thus, numerical stability is hard to achieve under these circumstances because finite-difference schemes locally may involve fewer than four surrounding cells. Nevertheless, while jetties add complexity and make simulations difficult, Experiment 8 suggests that WAVE can simulate sediment transport around jetties and is potentially useful for coastal engineering applications.

SIMULATING TRANSPORT ALONG THE OREGON COAST

Experiment 9 simulated a beach on the southwestern Oregon Coast described by Hunter and others (1979) and shown in Figures 1-1 and 2-9. The beach includes a longshore bar and rip channel, whose complicated submerged topography posed difficulties in computing wave refraction and wave-induced currents. Table 6-7 summarizes input data used in the experiment, which represented an area of 300 meters by 600 meters on a grid with 30 rows and 60 columns containing 1800 cells, each 10 meters square. At the outset the area was covered with a mixture of medium-grained quartz-feldspar sand having an average diameter of 0.5 millimeters.

Because of the complex nearshore topography, WAVE's circulation model WAVECIRC (Figure 3-15) was executed for 1500 iterations (requiring 15 minutes to compute on a Silicon Graphics Personal Iris workstation) to produce refraction and wave-induced circulation patterns shown in Figures 6-13 and 6-14. Currents shown by arrows in Figure 6-14 are appropriate given the complex topography, and compare well with observations of actual currents in Figure 2-9A. Currents in both Figures 6-14 and 2-9A moved offshore within the rip channel, shoreward over bar crests, and erratically along the shoreline. Simulated currents in Figure 6-14 suggest that vortices may occur where offshore and onshore currents mix.

The experiment's variable topography caused the surf zone to vary from 10 to 120 meters in width as waves generally broke in water depths of about 1.4 meters (Plates 1 and 14). As waves moved over the rip channel, greater water depths prevented them from breaking until they moved closer to shore, causing the surf zone to become narrow adjacent to the rip channel. By contrast, waves moving over bar crests broke farther offshore as they encountered shallow depths sooner, causing the surf zone to widen over bar crests. Transport rates along bar crests were as large as 0.01 cubic meters per second, but averaged 0.003 cubic meters per second within the entire surf zone. This range of simulated transport rates is within the range of rates observed for beaches of California and Florida (Figure 6-8).

161

Table 6-7 Experiment 9: Input data for beach on Oregon coast. Thickness of moving bedload and composition of beach sand were estimated from descriptions by Hunter and others (1979).

Wave parameters:		
Wave period (seconds)		10
Deepwater wave height (meters)		0.7
Deepwater wave angle (degrees)		230
Sediment transport:		
Maximum thickness of moving bedload (meters)		0.08
Coefficient K used in transport equation (Equation 4-17)		0.77
Grid parameters:		
Area of grid (meters)	300 x 600	
Grid cell length (meters)	10	
Number of rows i	30	
Number of columns j	60	

Sediment parameters:
Layer 1 (lower layer) at outset of experiment:
Thickness= 1.0 meter, average grain diameter = 0.5 mm:

	Diameter (mm)	Density (kg/m^3)	Percentage
Coarsest	1.50	2650	10
Medium	1.00	2650	10
Fine	0.50	2650	40
Finest	0.20	2650	40

Layer 2 (upper layer) at outset of experiment:
Thickness= 1.0 meter, average grain diameter = 1.0 mm:

	Diameter (mm)	Density (kg/m^3)	Percentage
Coarsest	1.50	2650	40
Medium	1.00	2650	40
Fine	0.50	2650	10
Finest	0.20	2650	10

Plate 1 shows that sand composition was also affected by the beach's complicated topography. For example, the greatest sorting occurred along bar crests, as shown by orange denoting coarser sand, amid greens and blues that denote medium and fine sand elsewhere in the surf zone. The enlarged fence display in Plate 15 shows differences in sediment composition along the bar crest, where the upper layers are coarse grained (yellow and orange), whereas medium-grained layers (green) were deposited away

162

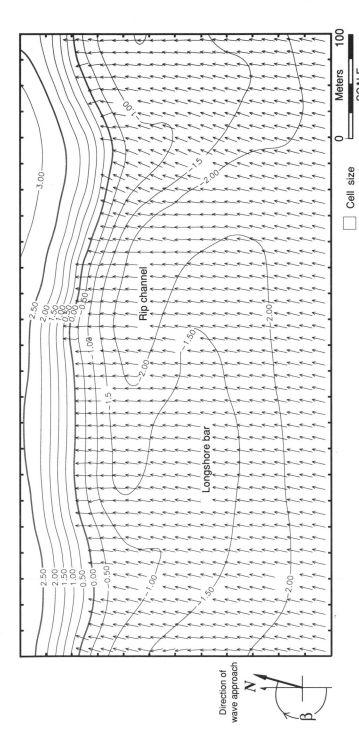

Figure 6-13 Experiment 9 - Beach on Oregon coast including longshore bar and rip channel. Wave orthogonals (arrows) were refracted by submerged topography. Deep-water wave angle was 230 degrees and wave height 0.7 meters. Contours show topographic elevations in meters with respect to mean sea level.

163

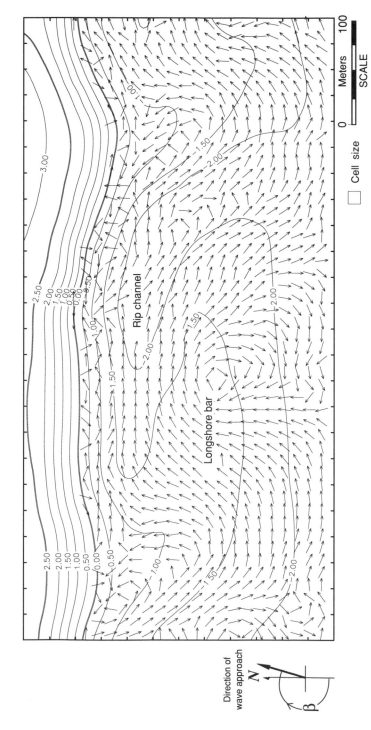

Figure 6-14 Experiment 9 -Beach on Oregon Coast: Directions of nearshore currents are shown by arrows. Currents flowed offshore within rip channels, and vortex occurred seaward of rip channel. Deep-water wave angle was 230 degrees and wave height 0.7 meters. Contours show topographic elevations in meters with respect to mean sea level.

from the bar. Layers in Plate 15 represent deposition over 24 successive intervals of time, each spanning 12 hours. Display of all 24 layers in rapid succession on a graphic workstation allows erosional and depositional events to be replayed in rapid motion, providing added insight into the processes that transport sediment alongshore.

CONCLUSIONS

Examples 3 through 9 demonstrate that WAVE can simulate beaches at varying geographic scales and can provide measurements of transport rates, sorting, wave heights, refraction angles, and other wave parameters. WAVE is potentially useful for forecasting effects of changing wave climates on beaches, and can simulate transport rates that appear to conform closely to those of actual beaches. If simulation of sediment transport is not a priority, WAVE can be used merely to provide one-time estimates of transport rates or wave climate, even in areas of complex topography, without involving extra iterations necessary to represent sediment tansport. Used in this way, WAVE is potentially useful for oceanographers and coastal engineers who prefer quick estimates of transport rates over more involved simulations. While static estimates of transport rates are valuable, WAVE can involve extra iterations to represent sediment transport, making it potentially useful for simulating iteractions between wave climate and beach topography. Simulations can also help forecast the long-term effects of jetties and other structures on longshore transport, while also providing a three-dimensional sedimentary record of the changes. WAVE's present disadvantage is that simulations involving long periods of time, ranging from hundreds to millions of years, require many hours of computer time. Faster computers in the future should allow computer procedures like those incorporated into WAVE to be used more effectively.

CHAPTER 7

Simulating
Longshore Transport
on Deltas

Experiments in Chapter 6 involving small areas and short spans of time are useful for comparison with field studies of modern beaches, but are less useful for interpreting ancient depositional systems that involve large areas and span hundreds or thousands of years. By linking WAVE and SEDSIM (Figure 7-1) we can perform experiments spanning longer times and larger areas that provide insight about sediment dispersal patterns and depositional history of both modern and ancient nearshore and deltaic deposits. Because deltaic deposits serve as important petroleum reservoirs, part of the discussion here focuses on WAVE's and SEDSIM's use as an exploration tool, where simulation experiments can provide insight about ancient buried deltas that are otherwise interpreted only from well data or seismic surveys.

We begin here with a brief review of SEDSIM and then describe Experiments 10 through 15 where WAVE and SEDSIM were linked to simulate deltas. Experiment 10 first simulated a delta without waves, whereas Experiment 11 involved a similar delta that was subjected to wave attack. The two experiments thus contrast a fluvial-dominated delta with a wave-dominated delta, and yield features that can be compared with actual deltas (Chapter 2). Experiment 12 involves two deltas that formed side by side that were affected by the combined influence of rising sea level and longshore transport by waves, thus documenting WAVE's and SEDSIM's versatility in representing a variety of nearshore processes that affect coastlines. Experiment 13 also involves two deltas affected by changes in sea level, but focuses on applications of WAVE and SEDSIM to sequence-stratigraphic analysis of ancient delta systems. Experiments 14 and 15 introduce "INTERACTIVE WAVE", a highly empirical or "parametric" model for simulating longshore transport. INTERACTIVE WAVE, as its name suggests, allows experiments to be controlled as simulations run, with "real-time" display of results.

167

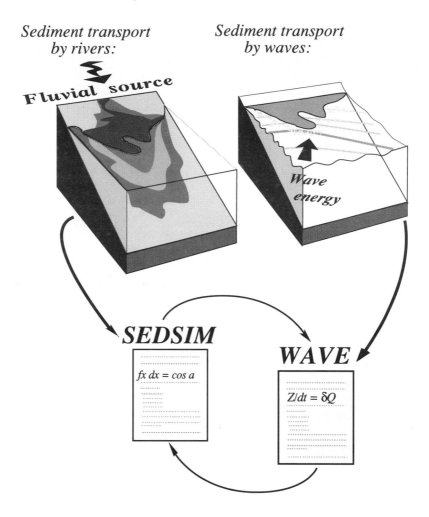

Sediment transport by rivers:

Sediment transport by waves:

Figure 7-1 Linkage of SEDSIM and WAVE for simulating sediment transport by rivers and waves.

SEDSIM

SEDSIM consists of FORTRAN 77 programs that incorporate laws of physics in simulating erosion, transport, and deposition of clastic sediment by flow in open channels. Tetzlaff and Harbaugh (1989) review SEDSIM's assumptions and governing equations. SEDSIM's objective is to improve understanding of sedimentary deposits formed in both modern or ancient fluvial and deltaic environments. Tetzlaff (1987), Scott (1987), Martinez (1987, 1992a, b), Martinez and Harbaugh (1989), Koltermann (1990a, b), Lee (1991), Lee and Harbaugh (1991), and Wendebourg (1991), used

SEDSIM to simulate depositional environments that include modern and ancient deltas, submarine fans, alluvial fans, braided streams, meandering rivers, and beaches.

SEDSIM'S representation of flow

SEDSIM's governing equations respond to input parameters such as fluid and sediment discharge rates that regulate the flow of water and sediment transport. Flow is represented with a marker-in-cell procedure where fluid elements move within a fixed grid (Figure 7-2). The marker-in-cell procedures allows flow velocities and sediment loads to be represented at points that move with the fluid elements, whereas parameters such as topographic elevations and water depths are stored at fixed positions within grid nodes. Each fluid element is a moving volume composed of water and suspended sediment that flows from a "source" (Figure 7-2A) such as a river mouth. Fluid elements move according to the Navier-Stokes equations for open-channel flow and obey laws of conservation of mass and momentum as they move and react to changes in topography and collide with other fluid elements. Fluid elements can have substantial volume, in contrast to the classical concept of fluid elements which have infinitesimal volume. Like WAVE, SEDSIM is quasi three-dimensional in its representation of flow. Both employ depth averaging in which flow velocities are averaged over the total water depth at all submerged grid locations.

Erosion, transportation, and deposition

Since fluid elements have volume, velocity, and momentum, they may erode, transport, or deposit sediment depending upon the local slope, their velocity, and the critical shear stress required to entrain sediment (Figure 7-3). In SEDSIM, a fluid element's ability to transport sediment is referred to as its "transport capacity" (Tetzlaff and Harbaugh, 1989). Transport capacities are continually updated as sediment is picked-up or deposited by a moving fluid element. As fluid elements move they react to changes in topography caused by erosion or deposition in previous time increments, creating a self-regulating system that incorporates feedback controls that mimic natural processes. Sediment transport equations used in SEDSIM are adapted from equations proposed by Meyer-Peter and Muller (1948) and described by Tetzlaff and Harbaugh (1989).

Like WAVE, SEDSIM can represent sediment composed of up to four grain types (Figure 7-4). Erosion, transport, and deposition are functions of fall velocity and critical shear stress, where coarser or denser grains are deposited first and finer or lighter grains later. The limitation to four grain types is a compromise between more realistic representation, and computing effort. More grain types could be accommodated with additional computational effort.

WAVE and SEDSIM simulate sorting in different ways because each employs different schemes for sediment transport. Figure 7-5 contrasts differences between their representation of flow and sedimentation. During an iteration, WAVE erodes sediment from a cell and transports it to directly adjacent neighboring cells, where it is immediately deposited (Figure 7-5B). With this scheme, sorting is necessarily represented during the erosion process, because once grains are eroded, they can only be

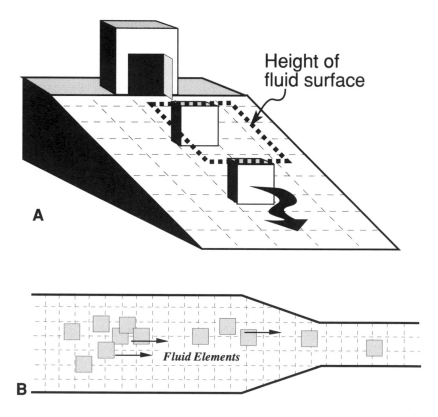

Figure 7-2 Schematic representation of marker-in-cell procedure for representing fluid flow: (A) Fluid elements (small blocks) emerge from fluid source (large block) and move in response to topography or differences in elevation of water surface (outlined by bold dashed lines) connecting upper surface of fluid elements. (B) Map view showing marker-in-cell procedure for representing moving fluid elements within fixed grid scheme. Flow velocities (arrows) are derived from fluid elements, whereas topographic elevations and water depth are defined at nodes of fixed grid denoted by dashed lines.

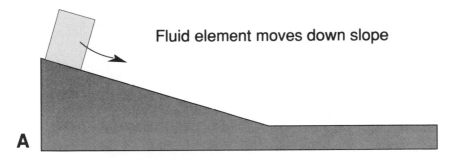

Fluid element moves down slope

A

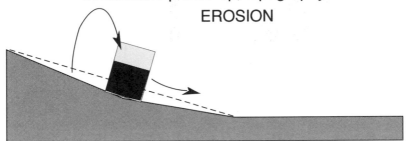

Velocity increases, shear stress increases, sediment is picked up, topography is lowered.
EROSION

B

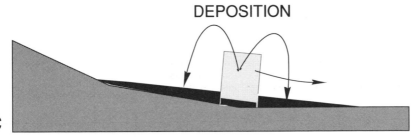

Velocity decreases, shear stress decreases, sediment is dropped, topography rises.
DEPOSITION

C

Figure 7-3 SEDSIM's procedure for representing sediment transport. (A) Fluid element moves down slope. (B) Increased velocity exceeds critical shear stress at sediment-water interface, eroding sediment and reducing elevation. (C) With decreased slope, velocity decreases and transport capacity is reduced so that sediment is deposited and elevation increased. After Tetzlaff and Harbaugh (1989).

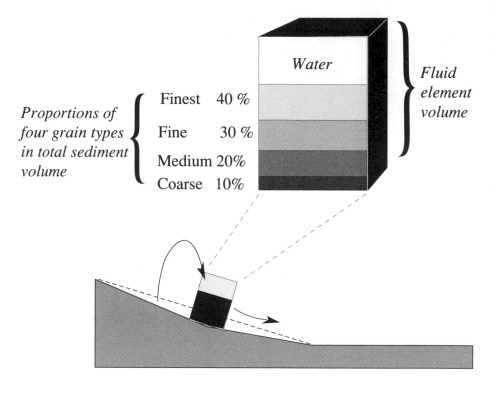

Figure 7-4 Diagram showing that total volume of fluid element includes water and sediment composed of up to four grain types, each having a specific grain density and diameter. Proportions of grain types vary as fluid element moves and deposition occurs in response to different fall velocities of grains.

moved as far as an adjacent cell, regardless of sizes or densities. Thus, WAVE controls sorting by *selective entrainment* (Figure 5-8A), where grain types are eroded according to critical threshold velocities and fall velocities. By contrast, SEDSIM's transport scheme allows fluid elements to move freely through a grid over successive time steps, so that sediment within fluid elements may be deposited either in immediately adjacent cells or in cells farther away (Figure 7-5A). Thus, SEDSIM simulates sorting by *settling equivalence* and *transport sorting* (Figure 5-8 B, C) by depositing grains as a function of fall velocity.

Extrapolation of time

Large-scale simulations such as those involving deltaic depositional environ-ments, may require representation of hundreds, thousands, or millions of years, but the processes that represent flow need to be represented at smaller time spans on the order of seconds, minutes, hours, or days. Thus, SEDSIM's computer procedures must represent processes occurring over a few seconds or hours, but then must employ

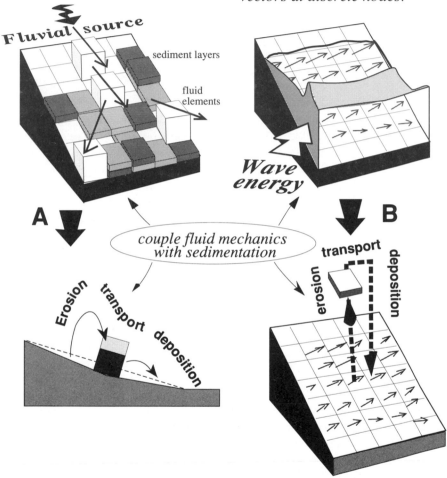

SEDSIM

represents open-channel flow with moving fluid elements:

WAVE

represents nearshore circulation with current-vectors at discrete nodes:

Fluvial source

sediment layers

fluid elements

A

Wave energy

B

couple fluid mechanics with sedimentation

Erosion transport deposition

transport

erosion deposition

Figure 7-5 Diagrams contrasting SEDSIM's and WAVE's representation of fluid flow and sediment transport: (A) In SEDSIM, fluid elements transport sediment as they flow freely over grid during successive time steps, permitting deposition in cells distant from those where sediment was eroded. (B) By contrast, WAVE moves sediment only to immediately adjacent cells during single time step, and new sediment transport rates are calculated during each time step.

173

enough time steps to represent hundreds or thousands of years. To reduce the number of iterations required to represent longer spans of time, SEDSIM incorporates extrapolation schemes where flow and sedimentation are assumed to be uniform over prescribed spans of time. For example, SEDSIM simulates changes in fluid flow and sedimentation by solving the Navier-Stokes equations for fluid motion for a few seconds, calculating the flow's effect on sedimentation, and then extrapolating the effects over much longer intervals of time spanning days, months, or years. Similarly, WAVE represents time by employing successive iterations, but the time steps employed by WAVE (Equations 5-28 and 6-1) are on the order of hours or days instead of seconds, allowing WAVE to simulate large intervals of time with acceptable computational effort, but without extrapolation.

Linking WAVE with SEDSIM

WAVE can function as a stand-alone program but it is implemented here as a subroutine within SEDSIM. As SEDSIM performs iterations over an interval of time *TE*, WAVE is called so that it simulates an equivalent interval (Figure 5-7). The interval can span a few hours to a few years. For example, if WAVE is called every 24 hours, it will simulate longshore transport over a 24-hour period before returning control to SEDSIM. Or, if WAVE is called every year, it will simulate longshore transport occurring during one year before returning to SEDSIM. Use of a relatively small time interval *TE* allows greater coupling or interaction between wave and fluvial processes, but requires more computational effort.

WAVE and SEDSIM share input data (Figure 7-6) involving initial topography and grain types, allowing arrays that contain topographic elevations and sediment layers (Figure 5-3) to be initialized within SEDSIM and then shared with WAVE. An example of WAVE's input file *wave.d* is shown in Table B-2 of Appendix B, whereas SEDSIM's file *input.d* is shown in Table D-1 of Appendix D. WAVE and SEDSIM also share subroutines that write data to output files, enabling each program to employ similar graphics procedures while avoiding redundant subroutines. Execution of SEDSIM and WAVE generally requires several hours of computer time, depending upon the amount of geologic time to be represented. For example, simulations of deltas such as those in Experiments 10, 11, and 12 that involve 600 to 1000 years, each required more than ten hours of computer time on a Silicon Graphics Personal Iris workstation. Furthermore, experiments involving both WAVE and SEDSIM may require 50 to 75 percent more computer time than experiments that either involve only SEDSIM, or only WAVE.

EXPERIMENT 10: DELTA WITHOUT WAVES

Experiment 10 simulates a delta without waves to provide a control experiment for comparison with a simulated delta in Experiment 11 that used similar input data, but included waves. Experiment 10 began with a uniformly sloping basement (Figure 7-7) whose initial slope was 0.011, or 0.63 degrees. The experiment employed

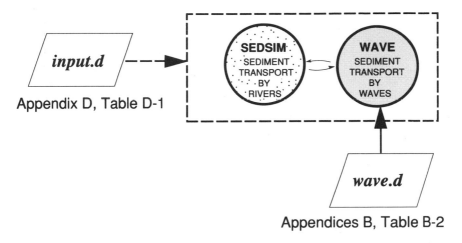

Figure 7-6 Diagram showing input files used by SEDSIM and WAVE. File *input.d* is shared by both, but *wave.d* is used only by WAVE.

a grid representing an area of 7.5 kilometers by 10 kilometers having 30 columns and 40 rows and containing a total of 1200 cells, each 250 meters square.

A single fluid source representing a river was located onshore near the northern edge of the grid (Figure 7-7). The source provided a steady flow of 250 cubic meters per second with fluid elements having initial velocities of one meter per second. The initial sediment concentration in fluid elements was 0.5 kilograms of sediment per cubic meter of water, corresponding to a sediment discharge rate of 125 kilograms per second. At the outset, sediment carried by each fluid element consisted of 10 percent fine sand, 20 percent very-fine sand, 50 percent silt, and 20 percent clay. Other input data are summarized in Table 7-1.

Experiment 10 involved 1000 years and required eight hours of computing time on a Silicon Graphics Personal Iris graphics workstation. After 1000 years, contours in Figure 7-8 show that the delta is generally symmetric and produced a bulge in the shoreline that extends approximately 1.5 kilometers beyond its original position (Figure 7-7). Plate 16 shows a three-dimensional view of the surface of the delta after 1000 years, where colors representing grain sizes at the water-sediment interface show that deposits are relatively symmetric around the river's mouth, with coarsest grains (orange) deposited near the river mouth and finer grains (green) deposited farther offshore.

Plates 17 shows a fence display after 1000 years, where orange, yellow, green and blue represent a continuum of grain sizes ranging from fine sand to clay. It is one of a series of fence displays that can be viewed in rapid succession with a graphics workstation to show the delta's evolution at 50-year intervals, recreating the delta graphically in video-like form. The fences compare well with general features of actual deltas, with coarser grains deposited near river mouths and finer sediments offshore.

175

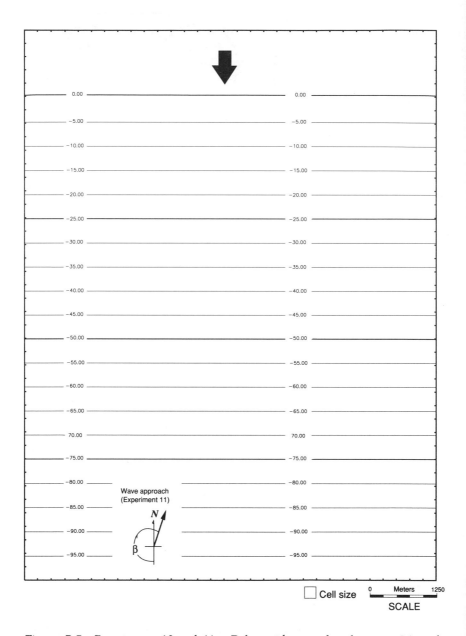

Figure 7-7 Experiments 10 and 11 - Deltas without and with waves: Map of topography at outset of both experiments, with contours in meters showing elevations with respect to mean sea level. Deep-water wave angle β in Experiment 11 is 200 degrees. Location of river source is shown by bold arrow.

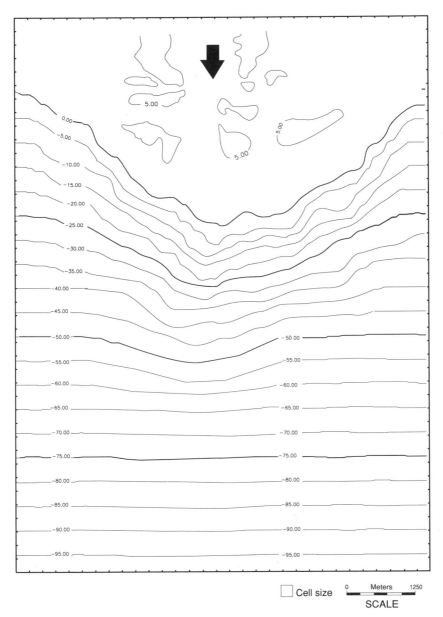

Figure 7-8 Experiment 10 - Delta without waves: Topography after 1000 years. Contours are in meters with respect to sea level. River source is shown by bold arrow.

177

Table 7-1 Experiment 10: Input data provided to SEDSIM for simulation of delta without waves.

Time parameters:

Simulated time (years)	1000

Fluvial sources:

Number of sources	1
Velocity of fluid elements (m/sec)	1
Fluid discharge (m^3/sec)	250
Sediment discharge (kg/sec)	125
Sediment concentration (kg/m^3)	0.5

Sediment parameters (four grain types):

	Diameter (mm)	Density (kg/m^3)	Percentage
Fine sand	0.10	2650	10
Very fine sand	0.075	2650	20
Coarse silt	0.025	2650	50
Fine silt	0.01	2650	20

Grid parameters:

Area of grid (meters)	7.5 km x 10 km
Grid cell length (meters)	250
Number of rows i	40
Number of columns j	30

Plate 18 is one of a series of fence displays showing ages of deposits. The series also can be displayed in video-like form, recreating the delta's history during the span of the simulation. Each band of color represents deposits formed during a 50-year interval, so that 20 such intervals represent 1000 years of geologic time. The fences show local discontinuities, particularly near the river's mouth and channel, where underlying beds were eroded as the river meandered.

Experiment 10 provides an example of a hypothetical delta formed by a single river that built a symmetrical prograding delta. Fluid and sediment discharge rates remained constant and effects of waves, tides, storms, sea level changes, subsidence, and other parameters affecting delta growth (Figure 1-9) were not included. Nevertheless, a realistic accumulation of sediment was obtained, suggesting that SEDSIM's marker-in-cell procedures for representing flow and sediment transport are useful for simulating deltaic systems.

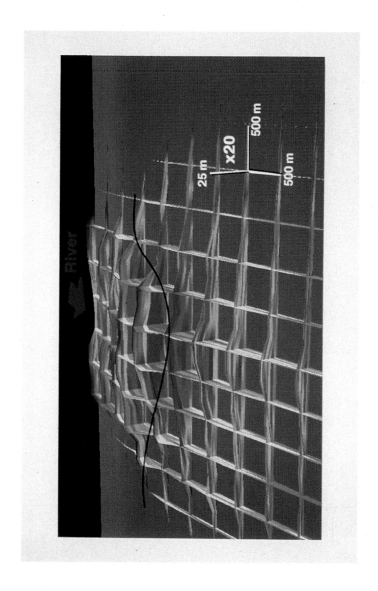

Plate 17 Experiment 10- Delta without waves. Perspective fence display of sections showing sediment composition. Orange, yellow, green, and blue represent range of grain sizes between fine sand, very-fine sand, coarse silt, and fine silt, respectively. Distance between sections is 500 meters. Vertical exaggeration is 20 times horizontal. Maximum thickness of sediment is about 30 meters.

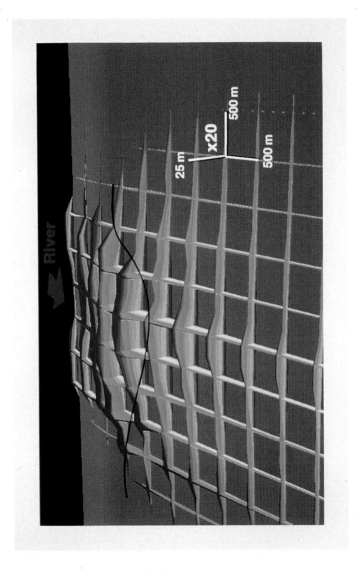

Plate 18 Experiment 10- Delta without waves. Fence display in which colors denote ages of layers. Each color band represents 50 years. Distance between sections is 500 meters. Vertical exaggeration is 20 times horizontal.

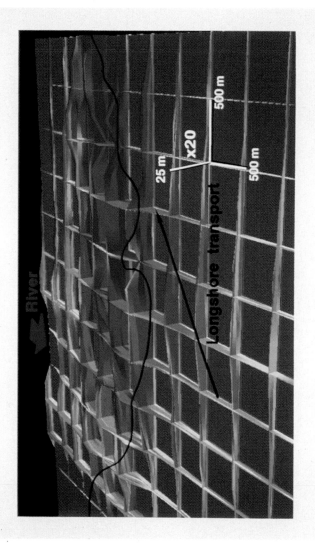

Plate 19 Experiment 11- Delta with waves. Fence display showing composition of layers. Orange, yellow, green, and blue represent range of grain sizes between fine sand, very-fine sand, coarse silt, and fine silt, respectively. Coarsest sand (red) was concentrated near shore towards right (east) of river mouth. Vertical scale is 20 times horizontal. Maximum thickness of sediment is about 30 meters. Distance between sections is 500 meters. Waves approached from lower left (southwest) corner.

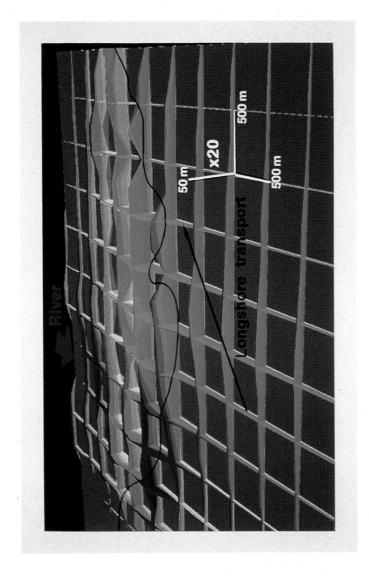

Plate 20 Experiment 11- Delta with waves. Fence display in which colors denote ages of layers. Each color band represents 50 years. Lobe-switching of deltaic deposits is. indicated by youngest sediments (red) that were deposited in west (left) part of area. Distance between sections is 500 meters. Vertical exaggeration is 20 times horizontal.

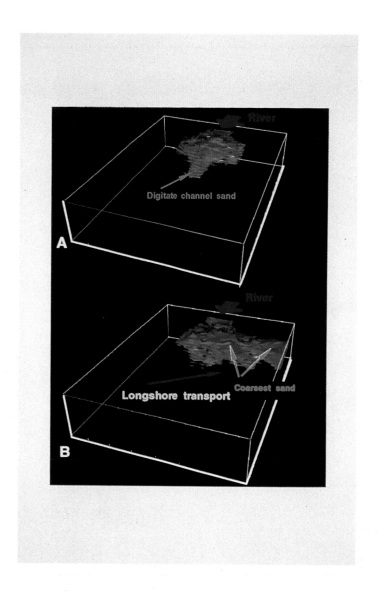

Plate 21 Experiments 10 and 11- Delta without waves and delta with waves. Perspective views show distributions of only those deposits whose average grain size is greater than 0.07 mm. Vertical scale is eight times horizontal. White lines denote box that is 7.5 km wide, 10 km long, and 200 meters tall: (A) Experiment 10 without waves: Deposits containing coarsest sediments are symmetric around river channel, but become digitate farther from shore. (B) Experiment 11 with waves: Deposits containing coarsest sediments are concentrated near shore and downdrift of river channel.

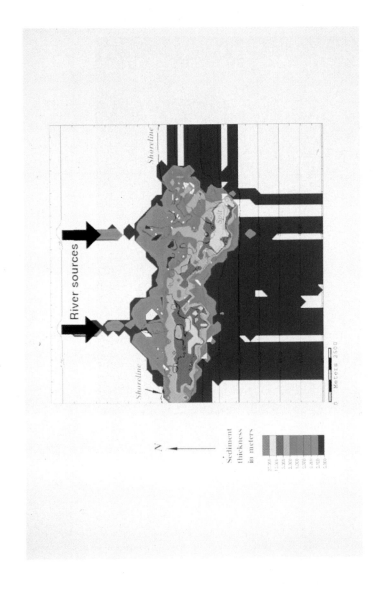

Plate 22 Experiment 12. Two deltas exposed to waves after 100 years. Map shows depths with contours in meters and thickness of sediments with colors. By this time, sea level had risen 1.5 meters. Thickest accumulations (yellow) occur at spit along front of easternmost delta. Bold black arrows show path of rivers, and bold black line shows position of shoreline. Waves approached from lower left (southwest) corner. Grid representing area has 50 rows and 50 columns containing 2500 grid cells, each 250 meters square.

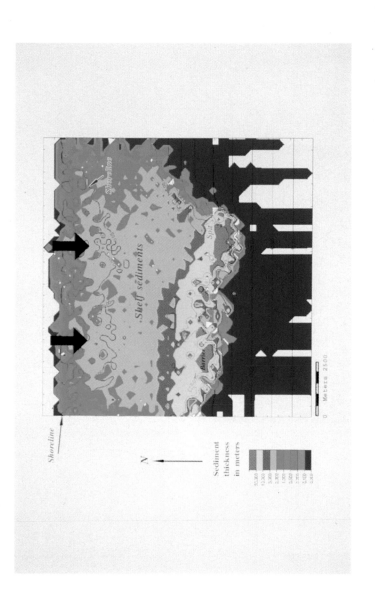

Plate 23 Experiment 12. Two deltas exposed to waves after 300 years. Map shows depths with contours in meters and thickness of sediments with colors. By this time, sea level had risen 4.5 meters. Thickest accumulations (yellow and light brown) occur within spits and barrier-bars near fronts of deltas, and thinner sediments (green) occur on shallow shelf Shoreline receded north during sea level rise.

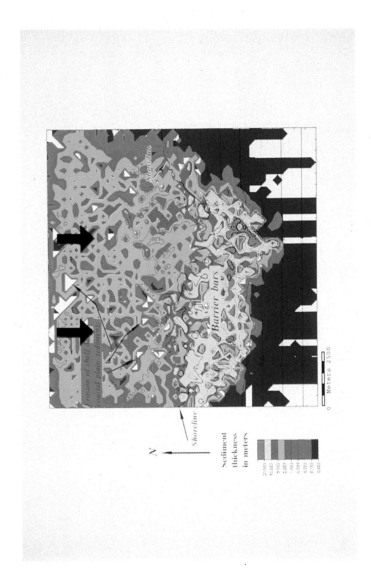

Plate 24 Experiment 12- Two deltas exposed to waves after 600 years. Map shows depths with contours in meters and thickness of sediments with colors. By this time, sea level had risen 7.5 meters, but remained unchanged between 500 and 600 years. Thickest accumulations (yellow and light brown) occur within spits and barrier-bars which became more prominent and are subaerially exposed. Thinner coastal plain and shelf sediments (green) deposited over shelf during sea level rise have been locally eroded in places (white) by meandering rivers. Shoreline moved south as system prograded during unchanging sea level between 500 and 600 years.

EXPERIMENT 11: DELTA WITH WAVES

Experiment 11 simulated a delta similar to that in Experiment 10, but included effects of waves. The two experiments permit a wave-dominated delta to be compared with a fluvial-dominated delta. Input data for Experiment 11 are identical to those for Experiment 10 (Table 7-1) except that additional data for WAVE are provided (Table 7-2). Waves approached from the lower left (southwest) corner of the grid, with deep-water wave heights of one meter and periods of ten seconds (Figure 7-7).

Experiment 11 involved 1000 years of geologic time, employed 20 successive displays to represent intervals of 50 years each, and required 18 hours to compute. Longshore currents ranged between 0.2 and 0.5 meters per second, and the average longshore transport rate was 0.006 cubic meters per second, or approximately 500 cubic meters per day. At 1000 years, contours in Figure 7-9 show that longshore transport by waves moved sediment to the right (east) and produced an asymmetric shoreline whose updrift topography is smooth and lobate in contrast to the irregular contours of the downdrift side.

Table 7-2 Experiment 11: Input data provided to WAVE for simulating delta with waves. Input for SEDSIM is in Table 7-1.

Wave properties:	
Interval *TE*, for calling WAVE (years)	1
Wave period (seconds)	10
Deepwater wave height (meters)	1
Deepwater wave angle (degrees)	200
Maximum thickness of moving bedload (meters)	0.05
Coefficient *K* used in transport equation (Equation 4-17)	0.77

Plate 3 is a three-dimensional view of Experiment 11 that shows sediment composition at the water-sediment interface at 1000 years. Sediment of different gain sizes is asymmetrically distributed around the river's mouth in response to longshore transport. Sorting in the surf zone of the coarsest grains (red, orange, and yellow) has moved them alongshore, whereas finer grains (green) remained in suspension to be deposited farther offshore, unaffected by longshore transport. The fence display in Plate 19 shows internal composition of the delta and emphasizes its asymmetric form and segregation of deposits caused by sorting and longshore transport.

Lobe switching is apparent in Plate 20 as the river's course moved persistently to the left (west) to avoid barriers in the form of spits and small bars that formed on the downdrift side of the delta. Bands of color representing ages of deposits show that they become progressively younger toward the west, with youngest deposits shown by red. Thus, waves caused changes in nearshore topography (Figure 7-9 and Plate 3) that influenced the river's course and affected dispersal and composition of delta deposits.

179

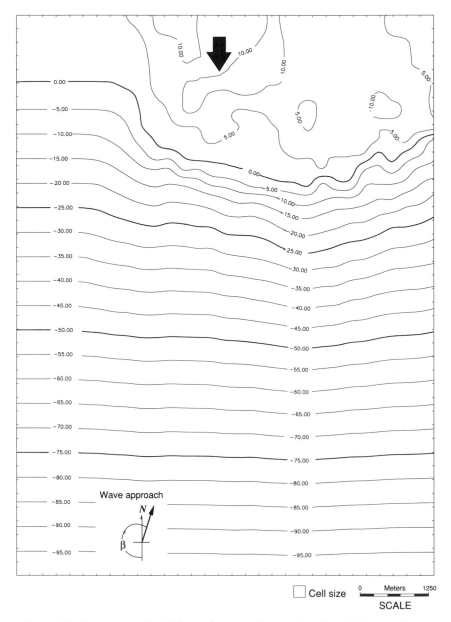

Figure 7-9 Experiment 11 - Delta with waves: Topography after 1000 years. Deep-water wave angle is 200 degrees.

Comparison with actual deltas

General features of actual fluvial-dominated and wave-dominated deltas outlined in Chapter 2 (Figures 2-13 and 2-14) can be compared with simulated deltas in Experiments 10 and 11. Differences in gross shape and composition of deposits in these experiments are contrasted in Plate 21, which shows deposits whose average grain diameter exceeds 0.07 mm. The delta in Experiment 10 has a digitate distributary channel containing coarse material that extends beyond the immediate shoreline (Plate 21A), similar to channel deposits in fluvial-dominated deltas (Figure 2-13) whose sands are generally thickest and coarsest in river channels near the shore. By contrast, Experiment 11 (Plates 19 and 21B) compares well with wave-dominated deltas (Figure 2-14) whose coarse sands are reworked and redeposited alongshore.

EXPERIMENT 12: DELTAS EXPOSED TO WAVES AND A RISING SEA

Experiment 12 involves two adjacent prograding deltas influenced by waves and rising sea level (Figure 7-10) in a setting similar to the typical coastline described in Chapter 1 (Figures 1-8 and 1-9). Experiment 12 employs a grid representing an area 12.5 by 12.5 kilometers and has 50 columns and 50 rows containing 2500 cells, each 250 meters square. At the outset the shoreline coincided with a change in slope denoted by closely-spaced contours in Figure 7-10B.

Fluid sources representing two rivers were located in channels in the upper half of the grid (large arrows in Figure 7-10B). Emerging fluid elements had an initial velocity of one meter per second (a discharge rate of 200 cubic meters of fluid per second) and contained suspended sediment of 0.4 kilograms per cubic meter of water (a sediment discharge rate of 80 kilograms per second). Sediment carried by fluid elements initially consisted of 10 percent fine sand, 20 percent very-fine sand, 30 percent coarse silt, and 40 percent fine silt. During the experiment, which spanned 600 years, sea level rose uniformly to 7.5 meters above its original position during the first 500 years and was unchanged during the last 100 years. Waves approached uniformly from the lower-left (southwest) corner of the grid, with deep water wave heights of 0.75 meters and periods of seven seconds, representing moderate wave conditions. Other input data are summarized in Table 7-3.

The experiment required 15 hours to compute and results are shown by Plates 22, 23, and 24, where colors show thickness of deposits after 100, 300, and 600 years of simulated time, respectively. Yellow and orange represent thicker accumulations, whereas blue and purple represent thinner deposits. After 100 years (Plate 22), spits (yellow) formed offshore of both deltas and were elongated east-west in the direction of longshore transport.

After 300 years (Plate 23), sea level rose 4.5 meters and flooded the gently sloping shelf that previously was above sea level (Plate 22) causing the shoreline to retreat northward and allowing two to five feet of fluvial and deltaic sediments to be deposited over most of shelf. Near the shoreline, waves reworked and transported older deltaic sediments to form spits and small barrier bars, similar to schematic representations in Figure 2-15.

181

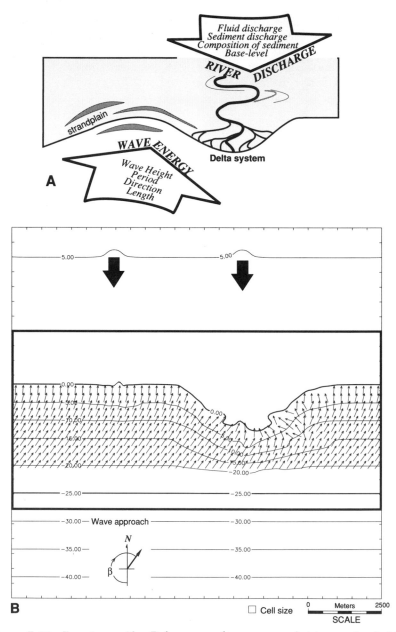

Figure 7-10 Experiment 12 - Deltas exposed to waves and rising sea level: (A) Schematic map of typical coastline introduced in Chapter 1, forming basis for Experiment 12, where river discharge rates and wave energy are provided as input. (B) Map of topography and wave orthogonals (small arrows) at outset of experiment. Contours in meters show elevations with respect to mean sea level. Locations of river sources are shown by bold arrows. Deep-water wave angle is 220 degrees. Box shows area represented in Plates 25, 26, and 27.

Table 7-3   Experiment 12: Input data used by SEDSIM and WAVE for simulating delta with waves and rising sea.

Time parameters:

Simulated time (years)   600

Fluvial sources:

Number of sources   2

Each source:

    Velocity of fluid elements (m/sec)   1

    Fluid discharge (m^3/sec)   200

    Sediment discharge rate (kg/sec)   80

    Sediment concentration (kg/m^3)   0.4

Sediment parameters (four grain types):

	Diameter (mm)	Density (kg/m^3)	Percentage
Fine sand	0.10	2650	10
Vy. fine sand	0.075	2650	20
Coarse silt	0.025	2650	30
Fine silt	0.01	2650	40

Grid parameters:

Area of grid (meters)	12 km x 12 km
Grid cell length (meters)	250
Number of rows i	50
Number of columns j	50

Sea-level change:

1.5 meters per 100 years, for first 500 years
Sea level held constant between 500 and 600 years

Wave properties:

Interval TE, for calling WAVE (years)	1
Wave period (seconds)	7
Deepwater wave height (meters)	0.75
Deepwater wave angle (degrees)	220
Maximum thickness of moving bedload (meters)	0.05
Coefficient K used in transport equation (Equation 4-17)	0.77

After 500 years of rising sea level, sea level stabilized at 7.5 meters above its initial position, but continuing sediment influx caused the shoreline to move southward (Plate 24) toward its original position (Figure 7-10B). Meandering river channels scoured and reworked underlying shelf sands on the shelf, stripping some areas of sediment (denoted by scattered areas of white amidst green and brown in Plate 24). Farther south, waves reworked older deltaic sands into small islands or barrier bars, similar to modern barrier bars and spits forming from the ancient Lafourche and St. Bernard deltas (Figure 2-11). These simulated bars and spits are transient features forming in response to the rise in sea level, and would probably be incorporated into terrestrial deposits with continued migration of the shoreline south. Similarly, many modern barrier bars and spits may be transient features resulting from recent fluctuations in sea level.

APPLICATION TO "SEQUENCE STRATIGRAPHY"- EXPERIMENT 13: SEA LEVEL CHANGES AS THEY AFFECT STRATIGRAPHIC SEQUENCES

Experiment 13 documents SEDSIM's ability to produce stratigraphic sequences that incorporate responses to major changes in sea level. Most petroleum geologists place strong emphasis on changes in sea level in interpreting oil and gas-bearing sequences. While sea level changes have had major effect in the past, their influence may be difficult to distinguish from other influences. For example, a shift in the mouth of a river as it enters a receiving basin may create deposits that are similar in sequence to those resulting from changes in sea level. WAVE and SEDSIM permit the effects of these alternative environmental influences to be explored when actual sequences are interpreted. Experiments involving sea level changes can be compared with experiments in which river volumes and source directions are changed, or with experiments that involve changes basin topography, subsidence, or wave climate. In this way, the effects of major parameters that affect sequences (Figure 7-11) may be explored in a series of experiments so that results of simulations can be compared to actual geologic cross sections or sections derived from reflection seismograms. These comparisons can provide better understanding of the factors affecting ancient or modern depositional systems.

Experiment 13 employs input data similar to Experiment 12 (Table 7-3) except that Experiment 13 spans 2000 years, incorporates a different history of sea level change, and for simplicity, does not include the effects of waves. Changes in sea level over the 2000-year span of the experiment are plotted in Figure 7-12. The first 500 years involved a straight-line fall of sea level of five meters with respect to its elevation at the outset of the experiment. Thereafter, sea level rose in straight-line fashion for the remaining 1500 years, so that at the end, sea level was 15 meters higher than the lowstand, and 10 meters higher than sea level at the outset. These rates of sea level

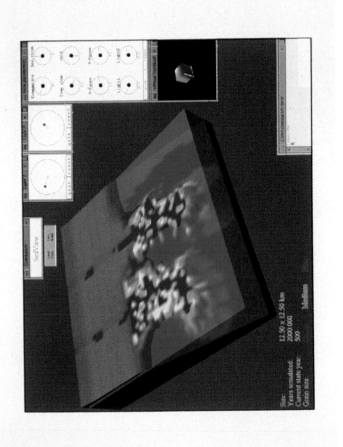

Plate 25 Experiment 13 involving fall followed by rise of sea level, 500 years after start of simulation: Perspective display shows two deltas formed by rivers that flow into receiving basin whose depth progressively increases seaward. Display shows composition of deposits at water-sediment interface, with colors denoting grain size. Colors represent a range of grain sizes between fine sand (red) through fine silt (blue). Sea level is shown by transparent light blue surface, and basement by grey. Display is controlled by adjusting dials using mouse or keyboard. Video-like display is obtained by rapidly showing results from a sequence of time steps. Vertical exaggeration is 40 times horizontal. Dimensions of block are 12.5 km by 12.5 km by 75 meters.

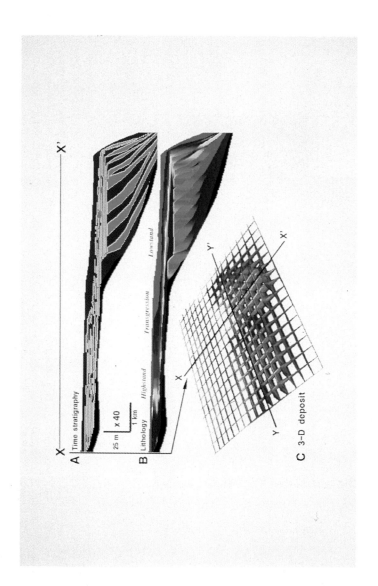

Plate 26 Experiment 13 involving fall followed by rise of sea level, 2000 years after start of simulation: Cross section X-X' is perpendicular to shore and has been isolated from perspective fence diagram. (A) Ages of deposits shown as alternating bands of red and blue, each band representing deposits formed during a 50-year interval. (B) Composition of deposits and sequence-stratigraphic interpretations. Colors represent range of grain sizes as in Plate 25. (C) Perspective fence display shows composition of beds denoted by color, and shows locations of section X-X' (this plate) and Y-Y (Plate 27). Vertical exaggeration is 40 times horizontal. Maximum thickness of deposits is about 55 meters.

Plate 27 Experiment 13: showing cross section Y-Y'. Sections (A) and (B) show ages and compositions of deposits similar to those of X-X' in Plate 26, except that Y-Y' is roughly parallel to shore. Buried channels are denoted by arrows.

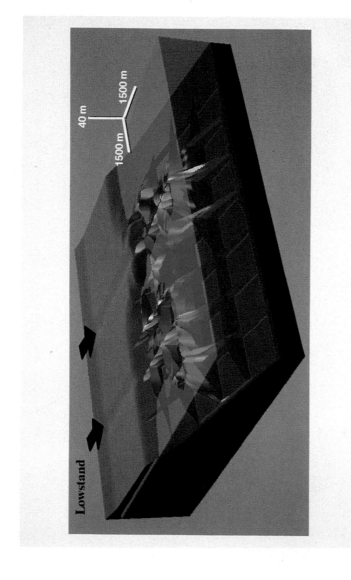

Plate 28 Experiment 13- Lowstand of sea level 500 years after start of simulation, when sea level was 5 meters lower than at start of experiment. Fence diagram shows sediment composition with colors denoting grain size, as in Plate 25. Sea level is transparent light blue surface. Arrows show paths of rivers. Vertical exaggeration is 40 times horizontal. Maximum thickness of deposits is about 40 meters. Distance between sections is 1500 meters. Dimensions of block are 12.5 km by 12.5 km by 75 m.

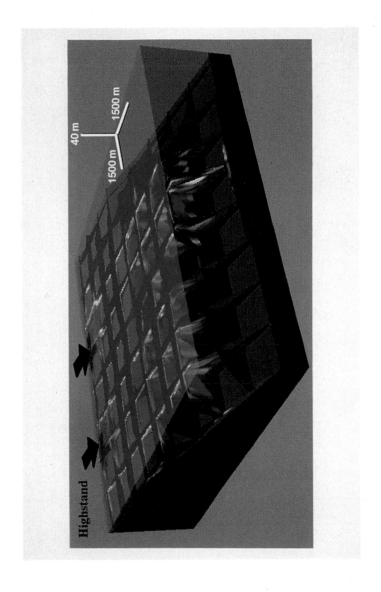

Plate 29 Experiment 13- Highstand of sea level 2000 years after start of simulation, when sea level was 10 meters above lowstand and 15 meters above sea level at start of experiment.

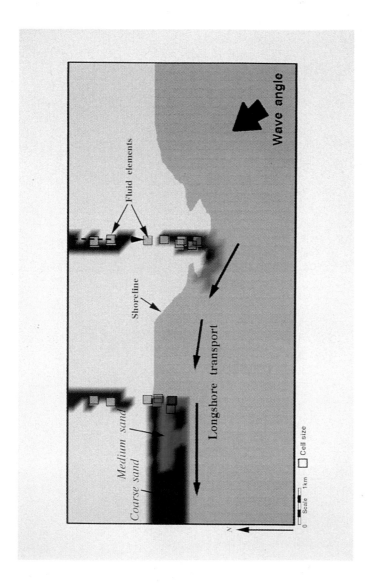

Plate 30 Experiment 14- Experiment with INTERACTIVE WAVE involving two deltas exposed to waves. Map shows river discharge and longshore transport after one simulated month, with composition of deposits shown by colors. Submerged offshore area shown by light blue, with wave approaching from lower right. Medium and fine sand (green and dark blue) from western river move alongshore over previously deposited coarse sand (red). Fluid elements are shown by boxes, with colors denoting their sediment composition. Dashed arrow shows path of fluid elements moving down channel from source. Bold black arrows show direction of longshore transport. Waves approached form lower right corner. Map is enlargement of area shown by box in Figure 7-10B.

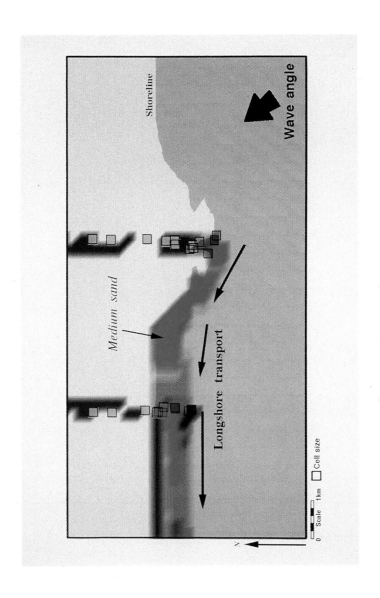

Plate 31 Experiment 14: Distribution of sediment after three months of simulated time. Medium and fine sand (green and dark blue) from eastern river preferentially move alongshore over coarser sand (orange).

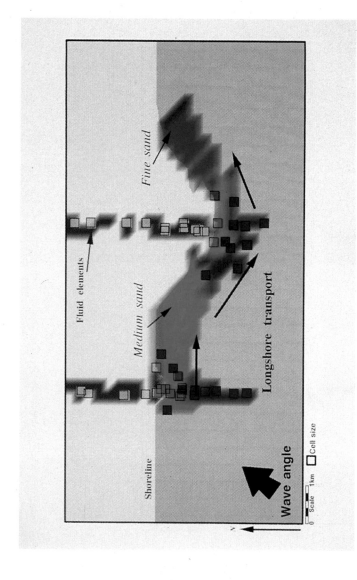

Plate 32 Experiment 15- Experiment with INTERACTIVE WAVE involving two deltas, with a reversal of the angle of wave approach: Map shows river discharge and longshore transport after one simulated year, with waves approaching from lower left corner. Composition of deposits are shown by colors. Medium sand (green) from western delta preferentially moves alongshore over finer sand (dark blue). Map is enlargement of boxed area in Figure 7-10B.

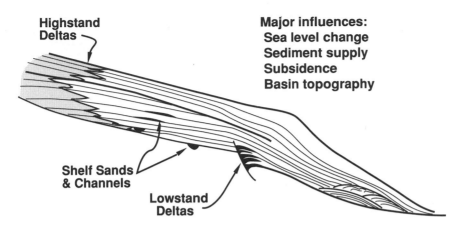

Highstand Deltas

Major influences:
Sea level change
Sediment supply
Subsidence
Basin topography

Shelf Sands & Channels

Lowstand Deltas

Figure 7-11 Cross section showing sequence stratigraphy of certain deltaic sequences of Texas and Louisiana. Section is a composite interpreted from reflection seismograms. Like Experiment 13, section reflects sequences formed during a drop in sea level, followed by a sea level rise. Lowstand, transgressive, and highstand deposits comparable to simulated deposits in Plates 25 through 29 are denoted by arrows. Major influences affecting sequences are sea level change, sediment supply, subsidence, and basin topography.

change are unrealistically large, but are employed to provide visually-obvious results for our discussion.

Plate 25 is a perspective view of two adjacent rivers entering a basin 500 years after the start of Experiment 13. While the two rivers have a constant discharge and flow independently of each other, the deltas at their mouths have begun to coalesce, so that the sequences of deposits within the deltas reflect not only their proximity to each other, but also the effects of local topography, and changes in sea level. Changing any one of these parameters would have a large influence on the sequence of simulated deposits.

Like previous experiments, we can graphically display simulated deposits at various times during the course of the simulation, so that the record of simulated deposits allows examination of features formed as sea level dropped and then later rose. We can show the record of simulated deposits by displaying fence-sections like that shown in Plates 26 and 27, to show both the ages and composition of simulated deposits at various times during the simulation. Sections showing ages of deposits (Plates 26A and 27A) can be compared to sections derived from reflection seismograms (Figure 7-11), which generally provide a record time-equivalent surfaces in the subsurface. By contrast, sections showing composition of deposits (Plates 26B and 27B) can be more directly compared to lithologic cross sections employing petrophysical logs from boreholes. Together, these two types of sections provide *time-stratigraphic* and *rock-stratigraphic* records, both of which are important to understand stratigraphy of sedimentary sequences.

185

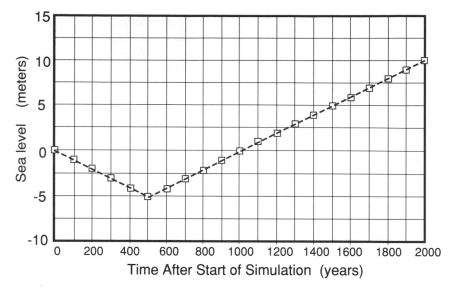

Figure 7-12 Graph of relative sea level change during Experiment 13.

Plate 28 shows sea level and the deposits after 500 years, at the "lowstand" of sea level. Regression of the sea has enabled the two rivers to flow across previous deposits, creating the succession of prograding foreslope deposits at progressively increasing distances from the former shoreline. Because both rivers flow into a steeply dipping basin with abundant accommodation space, deposition of delta-front sands is focussed near the shoreline, producing steeply-dipping foreslope deposits. These foreslope deposits (Plates 26) compare well with ancient deltaic deposits (Figure 7-11) formed during similar lowstands of sea level.

Plate 29 shows sea level and the deposits after 2000 years, at the "highstand" of sea level. With reversal of the change in sea level, rising sea level causes the river mouths to migrate shoreward as the sea transgresses, with large influence on deposits formed on the submerged shelf above the upper edges of previous foreslope deposits. Unlike lowstand deposits, highstand deposits are not focussed along a steep slope near the shoreline, but are widely distributed over the shallow shelf. Transgressive and highstand deposits have thin, vertically-stacked sequences, in contrast to steeply-dipping foreslope deposits of lowstand deposits (Plate 26).

Experiment 13 shows that a change in sea level combined with an irregular basin topography has a strong influence on the stratigraphy of simulated deposits. Beds formed during rising sea level are thinner and more widespread than foreslope deposits of lowstand deposits. While simple, the three-dimensional "package" of strata produced in Experiment 13 has features that are realistic and compare well in general form with actual sequences, such as ancient deltaic complexes in Texas and Louisiana (Figure 7-11). The comparisons support the concept that cyclic changes in sea level

186

strongly affected ancient stratigraphic sequences, and that individual "stratigraphic packages" formed during oscillations of sea level can be interpreted to yield estimates of the magnitude of the oscillations. The problem, however, is that specific sequences or packages can be produced under circumstances other than those that involve oscillations of sea level, so that interpretations are challenging.

Simulation models like WAVE and SEDSIM can provide help here because they allow investigators to perform any number of experiments where certain parameters are held constant, while others are changed, thereby allowing geologists to test their ideas concerning the influence of various parameters. For example, it would be useful to produce a series of simulated three-dimensional stratigraphic sequences classified according to the major environmental factors employed in their generation. These major factors could be varied systematically, so that geologists could select sequences obtained under specifically controlled environmental conditions for comparison with actual stratigraphic sequences. The series thus obtained could be organized with respect to rates of change in sea level, to changes in fluid and sediment discharge rates from rivers, to flow directions at river mouths, to prevailing wind directions and intensities, to responses of the crust to load, and to gross tectonic influences. Such a series should provide background for interpretations involving the subdiscipline of "sequence stratigraphy", which has received so much attention in recent years.

INTERACTIVE WAVE

Simulations involving WAVE and SEDSIM require a large computational effort which generally prohibits their use as "interactive" programs that can be graphically monitored and adjusted while simulations run. With greater computation power, results of simulations could be displayed as fast as they are computed, so that various input parameters could be continually adjusted and progress could be controlled much like a "flight simulator". With this goal in mind, a program called INTERACTIVE WAVE was developed and linked with a modified version of SEDSIM to provide interactive simulations of longshore transport, with "real-time" graphic display of results.

INTERACTIVE WAVE is an abbreviated procedure for computing sediment transport by waves. It is derived in part from the procedures previously described, but instead of using these relatively rigorous procedures, it employs a simplified scheme for representing longshore transport by waves, and as a consequence it runs much faster. In fact, INTERACTIVE WAVE runs so fast that users can change assumptions and observe the system's response almost immediately.

Development of INTERACTIVE WAVE departs from our previous effort to develop more rigorous procedures. Our discussion thus far has focussed on procedures for developing "process-response" simulation models where we attempt to mimic features of natural environments by first representing physical processes that form them. To simulate sediment transport on beaches, for example, we first developed procedures for representing shoaling waves and the currents that they produce. To represent deltaic deposition, we first developed procedures for representing open-channel flow. These "first-principle" approaches attempt to distill complex processes into fundamental physical principles that can be represented with mathematical expressions that adhere to the laws of conservation of mass, momentum, and energy. Process-response models tend to be more widely applicable than empirical or "parametric" models because they incorporate feed-back responses based on fundamental physical laws rather than a limited set of governing rules. Parametric models are defined here as models that represent the effects of physical processes, rather than trying to represent the processes themselves. They tend to incorporate empirical equations rather than equations based on fundamental physical laws. However, process-response models like WAVE and SEDSIM require a large computational effort that currently prohibits their use as interactive programs. We developed INTERACTIVE WAVE as a parametric model to demonstrate the advantage of interactive models that can be controlled as they run. As greater computing power becomes available, more rigorous procedures incorporated in WAVE can also be run as interactive programs.

INTERACTIVE WAVE does not simulate wave processes, rather, it simulates the movement of volumes of sediment which move according to transport rates, provided as input. Sediment moves only in cells that are submerged and whose water depths are less than wave base. This simplification eliminates the need to compute the location and characteristics of the breaker and surf zones. Furthermore, INTERACTIVE WAVE assumes that all sediment is moved alongshore, so that calculation of wave-induced currents is not required. The longshore direction is determined from the angle of wave approach provided as input, and the local beach slope. Wave angles are assumed to be the same in shallow water as in deep water, which results in a major simplification by eliminating the need to compute wave refraction. Table 7-4 that follows summarizes the main differences between INTERACTIVE WAVE and WAVE:

Table 7-4 Main differences between INTERACTIVE WAVE and WAVE.

INTERACTIVE WAVE	WAVE
(1) Nearshore currents NOT calculated: Sediment moves alongshore as a function of beach slope and angle of wave approach, provided as input.	(1) Nearshore currents are calculated: Sediment moves in direction of current.
(2) Refraction NOT calculated: Deep-water wave angle , provided as input, assumed to be equal to shallow-water angle.	(2) Refraction is calculated: Deep-water wave angle refracts and provides angle of incidence at breaker zone, for calculating transport rates.
(3) Position of breaker and surf zones is NOT calculated: Sediment moves in all submerged cell above wave base, provided as input.	3) Position of breaker and surf zone is calculated by computing behavior of shoaling waves: Sediment moves within surf zone, defined as area seaward of breaker zone.

(4) Transport rate NOT calculated:
Rates of sand transport provided as input.

(4) Sand transport rate is calculated from wave-power equation, that relates transport rate to wave energy and incidence angle at the breaker zone.

Of WAVE's many subroutines (Figure 5-7), INTERACTIVE WAVE employs only ZCHANGE, ERODE, SPLIT, MOVE, DPOSIT, and ZTNEW. These subroutines provide accounting schemes and other procedures for moving up to four grain types between cells. INTERACTIVE WAVE represents transport by calculating the volume of sediment that should be removed from cells based on transport rates, by the same procedures used by WAVE and described in Chapter 5, and in Figures 5-14 and 5-15. Similarly, the amount of sediment removed from cells is calculated from the continuity equation (Equation 5-27) used by WAVE. According to the continuity equation, the change in elevation of a cell during an interval of time is a function of the volume of sediment moving into or out of a cell, and the volume is obtained from the transport rate provided directly as input. Since transport rates are provided as input, INTERACTIVE WAVE merely sorts, moves, and distributes sand alongshore in those submerged cells that are above wave base. WAVE's procedures for representing differential movement of up to four grain types according to their grain diameters and densities are included, so that simulations can represent sorting by longshore transport.

INTERACTIVE WAVE provides a control panel on the computer screen (Figure 7-13) that allows the user to enter the following information using the keyboard or mouse, with changes in input reflected by the control panel:

Transport rate (m^3/sec): A single transport rate is supplied, and it pertains to all submerged cells that are above wave base. The rate selected pertains to rates on a cell by cell basis. Typical rates are provided in Appendix A. Rates of 0.001 to 0.01 cubic meters per second would be typical for experiments whose grid cells are 100 meters on each side.

Wave base (m) : Wave base is that depth above which sediment transport occurs and is generally defined as the depth where shoaling waves begin the feel the bottom and transport sediment. Wave base can be increased beyond typical depths to include offshore areas where processes other than waves are important, thereby allowing INTERACTIVE WAVE to represent transport by shelf or other currents. Wave base generally ranges between 0 and 20 meters, depending on the height of incoming waves.

Mobile bed depth (m) : Thickness of the mobile bedload (Figure 4-9) is assumed to be the same for all submerged cells above wave base, and is generally between 0.25 and 4 centimeters (Appendix A).

Wave call interval (years) : Wave call interval is the interval at which INTERACTIVE WAVE is called as a subroutine from SEDSIM. The interval may range from a fraction of a year to several years or more. Once called, INTERACTIVE

189

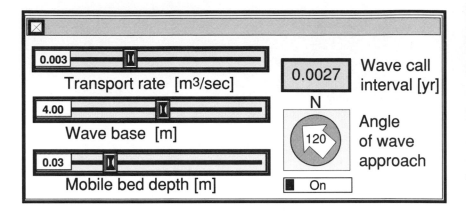

Figure 7-13 Display showing control panel that appears on-screen during simulation involving INTERACTIVE WAVE. Input data are adjusted while simulation runs by adjusting controls with mouse or entering data from keyboard. Angle of wave approach is changed by "grabbing" and rotating arrow. Transport rate, wave base, and depth of mobile bed are changed by sliding "scroll bars".

WAVE performs iterations until the interval of time specified has been represented, before returning control to SEDSIM.

Angle of wave approach (degrees) - Here, the user provides the direction of wave approach, which is the same as the deep-water wave angle.

EXPERIMENTS 14 AND 15 : LONGSHORE TRANSPORT BY INTERACTIVE WAVE

Experiments 14 and 15 illustrate INTERACTIVE WAVE's application. Experiment results are shown in Plates 30 and 31 for Experiment 14, and in Plate 32 for Experiment 15. Both experiments involve two rivers that are parallel and enter the receiving basin to create deltas a short distance apart. Input for Experiment 14 are shown in Figure 7-13. Incoming waves are from the southeast, so that longshore transport was toward the west. Colors in Plates 30, 31, and 32 show the composition of sediment types that were sorted and moved alongshore. Individual fluid elements are shown by colored boxes in river channels, as well as the nearshore area. As fluid elements erode or deposit sediment, their colors change to reflect the mixture of grain types they are carrying. The simulation is much like watching a movie, with fluid elements moving downslope depositing sediment at the shoreline, which is then transported alongshore as pulses or slugs of sediment of varying mixtures (colors) alongshore. Experiment 15 is similar to Experiment 14, except that the incoming waves are from the southwest, so that the direction of longshore transport is reversed. INTERACTIVE WAVE, while simple, demonstrates the advantage of dynamic, interactive simulation models, where input parameters and boundary conditions can be easily controlled, and the effects of these changes easily monitored. Faster computers will allow more complicated procedures like those in WAVE to be incorporated into interactive simulation models.

CONCLUSIONS

Experiments 12 through 15 conclude simulations described in this book by closely representing many features that occur on actual coastlines that were introduced as part of the "typical coastline" in Chapter 1 (Figures 1-8, 1-9, and 2-1). Experiments 1 through 15 provide examples of simulations that range from small beaches to large deltas, thereby documenting the geographic ranges over which WAVE may be applied. Experiments involving smaller areas can provide estimates of transport rates, nearshore wave heights, and wave angles, while experiments involving larger areas can provide insight about gross patterns of sediment dispersal, directions of longshore transport, and depositional history of deposits.

Simulations involving both large and small areas could be improved if WAVE were more flexible. For example, the small beach in Experiment 9 involving longshore bars in southern Oregon could be improved if wave-induced currents were represented more fully in three-dimensions. This would allow representation of both onshore-offshore and longshore transport by waves. Currently, WAVE does not fully represent fluctuating onshore and offshore transport caused by rip currents, return-flows, or oscillatory movements of waves because these fluctuating motions are time-averaged during a wave period and are provided as mean, depth-averaged values. Furthermore, simulations such as Experiment 12 that involve larger areas may represent longshore transport only within a single row of cells representing the surf zone (Figure 6-10), and could be improved by incorporating the effects of shelf currents or other currents that occur outside the surf zone, thus representing nearshore processes more completely.

Despite limitations, WAVE and SEDSIM incorporate generally accepted theories for representing water waves, open-channel flow, and sediment transport. When linked together they can simulate many aspects of nearshore environments and can provide a generalized set of experimental "tools" that permit variables to be selectively changed under controlled conditions. By contrast, neither physical scale models (such as wave tanks) or actual beaches can provide similar flexibility to control variables during experiments. WAVE, for example, can isolate the effects of a change in wave height while other variables remain unchanged whereas on an actual beach wave periods, wind speeds, and offshore wave heights constantly change, making it difficult or impossible to isolate effects of a single variable.

WAVE's major limitations are as follows:
(1) WAVE simulates sediment transport only in the surf zone, and neglects transport seaward of the breaker zone.

(2)   Sediment is transported only in the direction of time-averaged and depth-averaged currents and not in the direction of onshore or offshore oscillatory near-bottom currents.
      In actual situations, current directions may alternate during the passing of each wave.
(3)   Sediment transport rates are based on an empirical relationship between transport rates and wave heights observed on modern beaches, and is not directly a function of bottom shear stresses exerted by wave-induced currents.

These limitations could be reduced by incorporating procedures for representing wave-induced currents in three-dimensions. If near-bottom currents could be calculated efficiently, sediment could be transported as a function of bottom shear stresses caused by wave-induced currents. This would permit wave-induced currents to be coupled more rigorously with sediment transport, thereby avoiding WAVE's simplifications that involve two-dimensional representation of wave-induced currents and an empirical relationship for sediment transport. The computational effort of a more rigorous approach would be large, but as technology improves, the vast numbers of computations required to represent water waves, open channel flow, and sediment transport in three dimensions might be more feasible, allowing simulation experiments to run interactively, and to span thousands of years more effectively.

In this book we have tried to show that processes in nearshore environments can be represented by relationships derived from physics as well as by empirical relationships. These relationships can be represented by equations that can be solved with computers to carry out arithmetic and logic operations. Some of these relationships and underlying principles have application in geology beyond the nearshore environment, and although we have focused on applications to beaches and deltas, the procedures developed here can be applied elsewhere.

Perhaps most importantly, we have tried to show that nearshore processes act in concert with each other as part of the dynamic system represented by the earth as a whole. No process operates independently, and every process affects every other process to some degree. Geologists therefore need to interpret the earth as a dynamic system, and they need tools to do so. In our limited way, we have tried to provide some of these tools and to show how they can be linked so that alternative assumptions can be tested.

Observational data used to test WAVE

Table A-1 Averages of wave statistics of South Lake Worth Inlet, Florida based on observations during three-month interval described by Watts (1953). Location is shown in Figure 4-1. Longshore current listed here is average of 650 measurements. Wave height is average of 180 measurements. Wave heights and wave periods are daily measurements from wave gauge installed 20 km north of jetty at Palm Beach Pier in approximately five meters of water. Wave directions were measured with sighting bar and engineers transit from roof of nearby hotel. Longshore currents were measured by tracking speed of powdered dye released in surf zone.

	Longshore current	Wave height	Breaker angle	Period
	m/sec	meters	degrees	seconds
Number of measurements between 3- 52 to 6-52	(650)	(180)	(8)	(180)
Averages:	0.02	0.02	12	5

Table A-2   Longshore transport rates of South Lake Worth Inlet, Florida averaged from daily and monthly observations by Watts (1953). Sand transport rates estimated by measuring volumes of sand pumped through bypass plant near jetty. Measurements originally reported in English units have been converted to metric units. Numbers in bold type are averages of columns whose entries are themselves averages.

Date	Avgeraged over # days	Alongshore wave energy	Rate of sand transport	Rate of sand transport
month-year		watts/m, (or) newton-m/sec per meter of beach	m^3/day over cross section of surf zone	m^3/sec over cross section of surf zone
3-52	2.2	186.4	543	0.0063
3-52	2.0	183.3	249	0.0029
3-52	2.2	93.7	306	0.0035
3-52	0.5	231.7	985	0.0114
3-52	24.3	53.5	207	0.0024
4-52	1.3	57.1	217	0.0025
4-52	3.2	234.8	309	0.0036
4-52	2.0	130.5	261	0.0030
4-52	25.2	40.2	163	0.0019
5-52	4.2	83.4	251	0.0029
5-52	0.5	22.7	116	0.0013
5-52	0.3	74.1	173	0.0020
5-52	0.5	20.6	173	0.0020
5-52	30.8	18.0	58	0.0007
6-52	1.7	43.2	104	0.0012
6-52	9.2	9.5	43	0.0005
Grand averages:		**92.7**	**260**	**0.0030**

Table A-3 Average daily wave climate for selected months, at Anaheim Bay, California reported by Caldwell (1956). Location shown in Figure 4-3. Values are averages of observations spanning one month. Wave heights were measured using two types of wave gauges installed ten km south of jetty, at Huntington Beach Pier, at water depth of six meters. Significant wave heights, wave periods, and directions of wave approach were estimated from historical weather charts using hindcasting procedures. Measurements originally reported in English units have been converted to metric.

Date	Wave height	Breaker angle	Period
month-year	meters (@ 6 m depth)	degrees	seconds
10-48	0.8	0	14
11-48	0.3	5	15
1-49	0.3	5	16
3-49	0.5	5	16
4-49	0.4	5	15
Grand averages:	**0.5**	**4**	**15**

Table A-4 Longshore transport rates of Anaheim Bay, California based on observations by Caldwell (1956). Values are averages of observations spanning two or more months. Net rates of sediment transport were derived by measuring changes in topographic elevation along transects parallel to beach and calculating volumetric changes associated with topographic changes. Measurements originally reported in English units have been converted to metric units.

Date	Averaged over # days	Alongshore wave energy	Rate of sand transport	Rate of sand transport
month-year		watts/m, (or) newton-m/sec per meter of beach	m^3/day over cross section of surf zone	m^3/sec over cross section of surf zone
3-48	65	319.3	889	0.0103
6-48	66	916.8	1619	0.0187
8-48	95	123.6	230	0.0027
11-48	77	643.8	470	0.0054
1-49	73	545.9	1277	0.0148
4-49	123	87.6	642	0.0074
Grand averages:		**439.5**	**854**	**0.0099**

Table A-5 Average wave climate for selected days during months shown at El Moreno and Silver Strand beaches reported by Komar and Inman (1970). Location of beach shown in Figure 4-3. Table presents averages of values observed during experiments lasting several hours. Wave heights and periods were measured using digital amplitude staffs (DAS) and longshore currents were measured by releasing dye in surf zone.

Date	Longshore current	Breaker height	Breaker angle	Period
month-year	m/sec	meters	degrees	seconds
El Moreno:				
05-66	0.50	0.3	10	3
05-66	0.61	0.4	14	3
10-66	0.42	0.2	16	3
10-66	0.14	0.2	9	3
05-67	0.17	0.3	12	3
05-67	0.49	0.3	12	3
05-67	0.10	0.3	8	4
05-67	0.10	0.3	6	3
01-68	0.07	0.2	3	3
05-68	0.14	0.3	5	4
Grand averages:	**0.27**	**0.3**	**9**	**3**
Silver Strand:				
11-67	0.08	0.9	2	12
11-67	0.56	1.0	8	12
10-68	0.15	0.5	7	10
10-68	0.12	0.6	4	9
Grand averages:	**0.23**	**0.7**	**5**	**11**

Table A-6 Individual measurements from digital amplitude staffs (DAS) for El Moreno and Silver Strand beaches reported by Komar (1969) and Komar and Inman (1970). Measurements for wave height and angle represent values in deep water.

DAS observ. number	Period seconds	Height meters	Breaker height meters	Angle degrees	Breaker angle degrees	Orbital velocity m/sec
El Moreno:						
55	3	0.3	0.3	13	10	0.78
59	3	0.4	0.4	23	14	0.86
60	3	0.4	0.4	23	15	0.89
61	3	0.4	0.4	24	13	0.88
62	3	0.4	0.4	25	14	0.87
63	3	0.4	0.4	25	15	0.86
73	3	0.2	0.2	28	16	0.69
76	3	0.2	0.2	25	16	0.64
79	3	0.2	0.2	14	9	0.64
84	5	0.1	0.2	14	8	0.67
92	4	0.2	0.2	12	7	0.68
93	6	0.2	0.3	16	12	0.71
94	5	0.3	0.3	29	12	0.78
95	5	0.3	0.3	15	13	0.82
96	5	0.3	0.4	13	11	0.82
100	6	0.2	0.3	4	3	0.73
101	7	0.2	0.3	8	3	0.77
102	5	0.3	0.3	8	7	0.80
103	6	0.2	0.3	13	10	0.79
104	5	0.2	0.2	12	10	0.68
105	5	0.2	0.3	12	8	0.71
172	4	0.2	0.3	0	0	0.72
173	4	0.2	0.3	3	3	0.73
174	4	0.3	0.3	3	3	0.75
175	4	0.2	0.3	8	7	0.73
Averages:	4	0.3	0.3	15	9	0.76

(continued)

Table A-6 (continued)

Silver Strand:						
132	12	0.6	0.8	0	0	1.24
133	12	0.7	1.0	0	0	1.38
138	11	0.5	1.0	1	0	1.40
139	13	0.3	1.1	0	0	1.42
140	10	0.7	1.0	8	5	1.40
198	11	0.2	0.5	12	7	0.97
199	11	0.2	0.5	12	8	1.01
200	11	0.2	0.6	7	4	1.03
201	11	0.2	0.5	6	4	1.01
202	10	0.2	0.6	10	5	1.02
203	10	0.2	0.6	9	5	1.07
204	10	0.2	0.5	9	5	1.01
Averages:	**11**	**0.3**	**0.7**	**6**	**4**	**1.16**

Table A-7   Average longshore transport rates for selected days during months shown, at El Moreno and Silver Strand beaches reported by Komar (1969) and Komar and Inman (1970). Table presents averages of values observed during experiments typically lasting several hours. Rates of sand transport were estimated from movement of fluorescent sand tracers and estimates of thicknesses of moving bedload.

Date velocity	Grain wave energy	Alongshore transport	Rate of sand transport	Rate of sand
month-year	m/sec	watts/m, (or) newton-m/sec per meter of beach	m^3/day over cross section of surf zone	m^3/sec over cross section of surf zone
El Moreno:				
5-66	0.0050	43	388	0.0045
5-66	0.0065	104	724	0.0084
10-66	0.0014	30	221	0.0026
10-66	0.0007	15	85	0.0010
5-67	0.0008	20	125	0.0015
5-67	0.0022	38	368	0.0043
5-67	0.0004	6	52	0.0006
5-67	0.0006	18	76	0.0009
1-68	0.0003	6	25	0.0003
5-68	0.0014	18	178	0.0021
Grand averages:	**0.0019**	**30**	**225**	**0.0026**
Silver Strand:				
5-66	0.0010	15	110	0.0013
5-66	0.0055	380	2601	0.0301
10-68	0.0015	91	404	0.0047
10-68	0.0012	41	325	0.0038
Grand averages:	**0.0023**	**132**	**860**	**0.0100**

Table A-8 Observations of wave climate of five southern California beaches reported by Ingle (1966). Table presents averages of values observed during experiments lasting several hours. Wave heights were measured using wave staffs and wave directions from compass sightings.

Name of beach	Date	Longshore current	Breaker height	Breaker angle	Period
	mo-yr	m/sec	meters	degrees	seconds
Goleta:					
	7-61	0.15	1.1	9	9
	10-61	0.21	0.8	5	14
Trancas:					
	9-61	0.83	1.5	7	16
	9-61	0.34	0.9	4	11
	9-61	0.61	1.4	4	14
	11-61	0.28	1.1	9	11
	1-62	0.40	0.9	7	8
	3-62	0.28	0.7	4	12
	6-62	0.31	1.2	7	11
Santa Monica:					
	9-61	0.37	1.0	13	14
	12-61	0.92	1.1	4	11
	3-62	0.12	0.6	4	11
Huntington:					
	4-61	0.43	1.0	15	11
	9-61	0.52	1.4	14	11
	10-61	0.67	0.8	14	10
	1-62	0.21	1.1	0	15
	3-62	0.28	0.9	0	12
La Jolla:					
	6-61	0.40	0.9	20	14
	10-61	0.34	0.9	15	11
	11-61	0.15	0.9	2	13
	3-62	0.12	0.8	16	11
Averages for 5 beaches:					
		0.38	1.0	8	11

Table A-9 Longshore transport rates reported by Ingle (1966) for five southern California beaches. Rates of sand transport were estimated from movement of fluorescent sand tracers and estimates of thicknesses of moving bedload. Table presents averages of values observed during experiments lasting several hours, for selected day during month shown. Measurements originally reported in English units have been converted to metric units.

Beach	Date	Grain velocity	Alongshore wave energy	Rate of sand transport	Rate of sand transport
	mo-yr	m/sec	watts/m, (or) newton-m/sec per meter of beach	m^3/day over cross section of surf zone	m^3/sec over cross section of surf zone
Goleta:					
	4-61	0.088	703.0	2059	0.0238
	5-61	0.115	336.4	1078	0.0125
	7-61	0.050	338.6	69	0.0008
	9-61	0.038	126.8	160	0.0019
	12-61	0.054	136.5	841	0.0097
	1-62	0.075	210.9	536	0.0062
	3-62	0.078	185.6	342	0.0040
Trancas:					
	2-61	0.065	263.9	312	0.0036
	4-61	0.055	205.2	826	0.0096
	4-61	0.058	263.4	848	0.0098
	6-61	0.041	64.9	161	0.0019
	7-61	0.056	386.1	89	0.0010
	9-61	0.087	477.5	751	0.0087
	10-61	0.056	344.3	230	0.0027
	11-61	0.052	359.1	1270	0.0147

(continued)

202

Table A-9 (continued)

Santa Monica:				
2-61	0.044	314.6	1046	0.0121
4-61	0.054	355.5	782	0.0091
3-61	0.026	223.8	275	0.0032
7-61	0.033	53.5	57	0.0007
9-61	0.090	410.3	661	0.0077
1-62	0.028	63.2	239	0.0028
3-62	0.028	45.0	231	0.0027
Huntington:				
3-61	0.046	117.5	1066	0.0123
4-61	0.069	465.4	89	0.0010
6-61	0.060	343.5	1015	0.0117
7-61	0.063	333.8	158	0.0018
9-61	0.090	940.3	2185	0.0253
10-61	0.038	226.5	565	0.0065
1-62	0.116	981.6	182	0.0021
La Jolla:				
3-61	0.073	352.4	312	0.0036
4-61	0.076	444.1	299	0.0035
5-61	0.091	421.4	710	0.0082
7-61	0.083	355.5	277	0.0032
10-61	0.082	617.2	138	0.0016
11-61	0.040	47.2	125	0.0015
1-62	0.018	79.2	56	0.0007
Grand averages:	**0.052**	**269.6**	**466**	**0.0054**

Input data for WAVE's circulation module: WAVECIRC

Execution of WAVE's circulation model WAVECIRC requires file *wave.d* that provides data for variables described in Table B-1. Table B-2 provides an example listing of *wave.d*, which also includes data for controlling simulation of sediment transport. Typical ranges and suggested values for input parameters are described below. Martinez (1987b) reviews WAVECIRC and sensitivity analyses of input parameters. Successful simulations require that input values be confined to ranges discussed below:

Wave periods generally range from 1 to 20 seconds. During summer months average wave periods range between 3 and 12 seconds. Large storms offshore may generate wave periods of several minutes (U.S. Army Coastal Engineering Research Center, 1977), but tests with periods longer than 20 seconds have not been performed with WAVE. Mean annual wave periods for different regions of the United States are listed below (U.S. Army Coastal Engineering Research Center, 1977, p 4-32) as well as in Appendix A.

Gulf Coast	3.5 seconds
Atlantic Coast (south)	4.8 seconds
Atlantic Coast (north)	6.5 seconds
Oregon- Washington	9.0 seconds
California	11.0 seconds

Deep-water wave heights generally range from 0.3 to two meters. While storms can generate deep-water wave heights as large as ten meters, experiments involving wave heights larger than three meters have not been performed with WAVE. Some mean annual deep-water wave heights along coasts of the United States are listed

Table B-1 Input variables for WAVE's circulation module, WAVECIRC.

Variable name	Description
T	Wave period (seconds)
H	Wave height in deep water (meters)
A	Wave angle in deep water (degrees)
WIND	Wind speed (m/sec)
WINANG	Wind angle (degrees)
ITA	Total number of iterations used in numerical scheme to achive steady-state currents
NHIGHT	Number of iterations to build-up deepwater wave height
KSKIP	Print output every *n*th iteration
MAXDEP	Maximum water depth to simulate waves (depth wave base)
MNODE, NNODE	Row and column numbers to write x and y components of currents
M	Number of rows in grid
N	Number of columns in grid
DX	Length and width of grid cells in meters

below (U.S. Army Coastal Engineering Research Center, 1977, p 4-33) and in Appendix A.

Atlantic Coast	0.62 meters
Gulf Coast	0.50 meters
Pacific Coast	0.98 meters

Wave angles must be between 90 and 270 degrees measured clockwise, from the +x axis, as shown in Figure 3-13. WAVECIRC's procedures assume that shorelines trend generally east-west and are located near the northern edge of the grid (Figures 3-12 and 3-13).

Wind speeds typically range from 0 to 40 knots, but speeds larger than 20 knots have not been tested with WAVE. WAVE requires wind speeds expressed in meters per second, where one knot equals 0.5144 meters per second.

Wind angles are measured in degrees clockwise from the *x* axis, as shown in Figure 3-13.

M and N specify grid size (Figure 3-12) and can be as large as the array dimensions defined in WAVE and SEDSIM, which are 50 rows and 50 columns presently. Increasing **M** or **N** increases the number of cells in the grid, thereby establishing the

Table B-2 Actual listing of WAVE's input file "wave.d" which is edited by user to provide input data for controlling wave climate and sediment transport. User may change numbers in file with text-editor program. Variable names are ignored when file is read by WAVE. Table B-1 defines input variables for WAVE's circulation module WAVECIRC, whereas Table C-1 defines input required to control sediment transport.

```
**********************************************************************
       INPUT DATA FOR SIMULATION SEDIMENT TRANSPORT BY WAVES
**********************************************************************
```

INPUT FOR CIRCULATION MODULE "WAVECIRC":

T:	7.0
H:	0.5
A:	190.0
WIND:	2.5
WINANG:	60.0
ITA:	1000
NHIGHT:	200
KSKIP:	1000
MAXDEP:	10.0
MNODE:	5
NNODE:	5
M:	50
N:	50
DX:	10

INPUT FOR CONTROLLING SEDIMENT TRANSPORT MODULE:

DMOBILE:	0.08
XKK:	0.77
ICALL:	1
LIMIT:	500

```
**********************************************************************
```

geographic area represented by the grid whose cells have dimensions DX and DY, described below.

DX and DY represent grid spacings Δx and Δy shown in (Figure 3-12) and, along with M and N (above), establish the scale of the simulated area. DX and DY are assumed to be equal. In experiments, DX and DY have ranged from a few meters to hundreds of meters. Selection of DX and DY depends upon the detail desired, with accuracy increased with smaller DX and DY.

207

Time step DT is used for obtaining numerical solutions with the finite-difference scheme of Equation 3-76. DT must satisfy the following criterion based on a Courant criterion proposed by Birkemeier and Dalrymple (1975):

$$DT \leq DX / (2g\,d_{max})^{1/2} \tag{B-1}$$

where :

DT = time step Δt in seconds defined in Equation 3-75
DX = length or width of grid cell Δx given as input (Figure 3-13)
           Equation B-1 assumes that DX = DY
d_{max} = maximum depth encountered in grid (known at outset of experiment, as established by initial topographic elevations defining simulation area).
g = gravitational constant

With Equation B-1, a maximum value for DT can be determined if cell size DX, and maximum water depth dma_x are given. For convenience, DT is calculated automatically within WAVECIRC using Equation B-1, although it could be provided directly as input. When DT is provided as input, it must be less than or equal to the right side of Equation B-1. Equation B-1 predicts that WAVE's numerical solutions will become unstable if DT is comparatively small when used with a grid whose cell size DX is large. Inspection of Equation B-1 shows that the resolution represented by the grid is governed by DX and DY, which in turn govern the selection of DT.

ITA specifies the total number of iterations during a single run of WAVE. Generally, WAVE should be run for several hundred iterations so that steady state solutions of wave-induced currents are obtained. WAVE's numerical solutions reach a steady state when wave setup and current velocities exhibit little change during consecutive iterations (Figures 3-17, 3-18, and 3-19).

NHIGHT specifies the number of iterations used to build up deep-water wave height to full value. Gradual buildup avoids "shocking" the system and allows procedures that represent longshore currents to converge.

KSKIP regulates the frequency that data are printed as an output file, and must be less than or equal to ITA, the total number of iterations during a single run of WAVE.

MAXDEP specifies maximum water depths where WAVE's procedures operate, thereby representing wave base. It avoids unnecessary calculations in deep water where waves do not affect the bottom.

APPENDIX C

Input data for WAVE's sedimentation module

Execution of WAVE's sedimentation module (organization shown in Figure 5-7) requires data file *wave.d* (Table B-2) that provides input variables listed in Table C-1 and described below.

DMOBILE (Figure 4-9) represents the thickness of moving bedload described and measured by King (1951), Komar and Inman (1970), and Madsen (1989). Its may range from millimeters to several centimeters. Experiments with WAVE have employed thicknesses ranging between 0.01 and 0.1 meters.

XKK is the dimensionless coefficient K (Figure 4-2) used in solution of Equation 4-17 for longshore transport rates. K ranges between 0.6 and 1.0 (Table 4-1). Komar (1988) reviews its calibration and techniques for estimating K from field observations.

ICALL determines frequency that WAVECIRC is called during iterations of WAVE's sedimentation module. Wave climate may be recalculated by WAVECIRC during each cycle of the sedimentation module (Figure 5-7), but computing time is saved by recalculating it less often by using velocities obtained in previous time steps, permitting WAVE's sedimentation model to return to subroutine DIRECT (Figure 5-7) rather than WAVECIRC. Experiments presented here all involved setting ICALL set to one, so that wave climate was recalculated during each iteration. Use of values greater than one for ICALL implies that wave climates are valid for subsequent iterations and involves extrapolations where sensitivity has not yet been investigated.

LIMIT limits the maximum number of iterations allowed during any single call to WAVE (Figure 5-7), a precaution that is necessary because WAVE's sedimentation model employs a time step that is recalculated during each iteration (Equation 5-28), thereby changing the number of iterations performed during each call to WAVE. An

Table C-1 Input variables required by WAVE's sedimentation module.

Variable	Description
DMOBILE	Maximum thickness allowed for depth of mobile bed (meters)
XKK	Coefficient K used in sediment transport equation (Equation 4-17)
ICALL	Recalculate wave climate after every nth iteration
LIMIT	Maximum number of iterations allowed, each call to WAVE

appropriate selection for LIMIT can be obtained by executing WAVE for a few minutes, observing the number of iterations required during each call to WAVE, stopping the program, and setting LIMIT to a similar but slightly higher number.

Additional information is required to describe the composition of sediment layers present at the outset of each experiment. For example, arrays such as those shown in Figure 5-3 are used to record ages and compositions of sediment cells and are important for WAVE's procedures that erode and transport sediment. They provide information needed by both SEDSIM and WAVE via arrays shared by both WAVE and SEDSIM, with some information used by WAVE provided through SEDSIM, and not from an input file. Information for grain sizes, densities, and thicknesses required by WAVE are not supplied in WAVE's input file *wave.d*, but instead are provided by SEDSIM through its input file *input.d* described in Table D-1 and shown schematically in Figure 7-6.

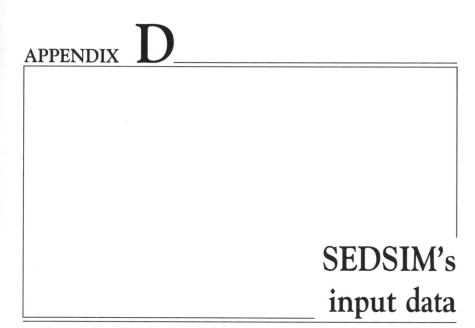

SEDSIM's input data

Example input data for SEDSIM are shown in Table D-1. Descriptions and suggested ranges of variables are described below and by Tetzlaff and Harbaugh (1989).

RUN-TIME PARAMETERS

The Navier-Stokes equations for representing open-channel flow are time dependent, as are equations for sediment transport rates. Thus, representation of time is important in a dynamic time-dependent model. Flow equations and sediment transport rates must be computed at intervals spanning several seconds of time, but the geologic processes being simulated operate over spans of days, years, or thousands of years. Run-time parameters control the amounts of geologic time to be simulated and the intervals at which graphic output is to be written to a graphics file for display. In SEDSIM's Fortran code, the variables are:

TO = START TIME of simulation, in years
TR = END TIME of simulation in years
TD = TIME/DISP, or display time interval, in years

START TIME is generally 0.0 years, unless output from a previous SEDSIM experiment is to be used as input to begin an extended simulation.

END TIME is the total number of years to be simulated. In the example of Table D-1, SEDSIM will simulate 10,000 years of geologic time. The amount of computer

Table D-1 Example listing of SEDSIM's input data file "input.d".

TITLE/DESCRIPTION: Put your title here, up to 72 characters

Run Parameters:
START TIME	(years) =	0.00
END TIME	(years) =	10000.00
TIME/DISPL	(years) =	500.00

Extrapolation Factors:
DURATION OF FLOW (seconds)	=	10.00
SED. EXTRAPOLATION (years)	=	1.00
SEDIMENT EXTRAPOLATION FACTOR	=	10.00
SOURCE INTERVAL (years)	=	1.00

Physical Constants:
FLOW DENS. (kg/m3) = 1027.00
SEA DENS.   (kg/m3) = 1013.00

Sediment Parameters:
COARSE	MEDIUM	FINE	FINEST	
0.001	0.0005	0.00025	0.000125	DIAMETER (meters)
2650.00	2650.00	2650.00	2650.00	DENSITY (kg/m3)
0.20	0.40	0.20	0.20	BASEMENT (ratios)

Fluvial Sources:
NUMBER OF SOURCES = 2
X(m)	Y(m)	Xveloc(m/s)	Yveloc(m/s)	Discharge(m3/s)
300.00	700.00	0.00	-0.50	50.0
400.00	800.00	0.00	-0.50	100.0

Sediment load for each fluvial source
(%)COARSE	(%)MEDIUM	(%)FINE	(%)FINEST	CONCENTRATION (kg/m3)
50.00	20.00	20.00	10.00	0.500
10.00	20.00	30.00	40.00	0.800

Topography:
GRID SIDE (m) = 100.0
NROWS   = 9.0
NCOLS   = 8.0

GRID NODES ELEVATION (BASEMENT) (m)

5.00	5.00	5.00	5.00	5.00	5.00	5.00	5.00
3.75	3.75	3.75	3.75	3.75	3.75	3.75	3.75
2.50	2.50	2.50	2.50	2.50	2.50	2.50	2.50
1.25	1.25	1.25	1.25	1.25	1.25	1.25	1.25
0.00	0.00	0.00	0.00	0.00	0.00	0.00	0.00
-1.25	-1.25	-1.25	-1.25	-1.25	-1.25	-1.25	-1.25
-2.50	-2.50	-2.50	-2.50	-2.50	-2.50	-2.50	-2.50
-3.75	-3.75	-3.75	-3.75	-3.75	-3.75	-3.75	-3.75
-5.00	-5.00	-5.00	-5.00	-5.00	-5.00	-5.00	-5.00
"	"	"	remaining values not shown			"	"

(CPU) time required to run the experiment will depend on various factors, including the amounts of extrapolation used in the simulation and number of fluvial sources within the simulated coastline.

TIME DISPLAY is the interval at which graphic output is written to a graphics file for display. In the example in Table D-1, a graphics file is to be written every 500 years, or 20 times (10000 ÷ 500 = 20) during the experiment. Graphics files are so large that memory limitations currently constrain the number of displays generated to range from 20 to 50, depending upon their size, and TIME DISPLAY should be chosen accordingly.

EXTRAPOLATION

Shortcuts must be employed to reduce computational labor. For example, END TIME in the example input of Table D-1 is 10,000 years, but the actual simulation required only a few hours of computer time because it involved extrapolation. Thus, if we wish to simulate significant amounts of geologic time, extrapolation is necessary. Extrapolation is accomplished in SEDSIM by making accurate calculations for short intervals of time and extrapolating their effects over additional intervals of time, as prescribed in the input. Extrapolation is the least intuitive of SEDSIM's parameters, and its use is described further by Tetzlaff and Harbaugh (1989).

PHYSICAL CONSTANTS

Physical constants determine whether river water flowing from a fluid "source" will be more dense or less dense than sea water into which it flows.

FLOW DENSITY is the water density of each fluid source. Fresh water typically has a density of 1000.00 kg/m^3.

SEA DENSITY is the density of water not derived from fluid sources. Any location within the grid below sea level will have water with this density. Seawater typically has an average salinity of 35,000 ppm and a density of 1146.00 kg/m^3. Of course, the body of water could also be a lake containing fresh water.

MULTIPLE GRAIN TYPES

Up to four types of clastic sediment can be defined. Combinations of the four end member grain types are represented by different colors in graphic displays, where COARSE=red, MEDIUM=yellow, FINE=green, FINEST=blue, and mixtures of these end members are represented by mixtures of the corresponding end-member colors.

DIAMETER in mm is the grain diameter of each grain type. Table D-2 lists grain diameters in the Wentworth classification. SEDSIM's empirical equations for erosion, transport, and deposition are best suited for sand, silt, and clay-sized particles.

DENSITY in kg/m^3 defines the grain density of each grain types. The user can provide densities equivalent to quartz, carbonate, or heavy minerals. Densities of some rocks and minerals are listed in Table D-3.

BASEMENT ratios define the composition of the basement as proportions of the four clastic grain types as previously defined. Material eroded from the basement provides sediment in proportions provided by the ratios, which must sum to 1.0

Table D-2 Wentworth grain-size and phi classification scale. Adapted from U. S. Army CERC (1977).

Grain or clast	Grain diameter (mm)	PHI ϕ designation
Boulder		
	256	-6
Cobble		
	64	-4
Pebble		
	4	-2
Granule		
	2	-1
Very coarse sand		
	1	0
Coarse sand		
	0.5	1
Medium sand		
	0.25	2
Fine sand		
	0.125	3
Very fine sand		
	0.0625	4
Silt		
	0.00391	8
Clay		
	0.00024	12
Colloid		

Table D-3 Densities of some rocks and minerals. Adapted from
U. S. Army CERC (1977).

Rock/Mineral	Density (kg/m^3)
Quartz	2650
Limestone	2710
Dolomite	2876
Shale, clay	2200 - 2750
Lignite	700 - 1500
Muscovite	2760 - 2880
Coal	1300 - 1500
Gold	15000 - 19300
Tourmaline	3000 - 3250
Zircon	4680
Magnetite	5180
Pyrite	5020
Rutile	4180 - 4250
Corundum	4020
Garnet	3500 - 4300
Pyroxene	3200 - 3500
Amphibole	3000 - 3400
Feldspar	2540 - 2570

FLUID SOURCES

Fluid elements emerge from fluid "sources", whose locations in the simulation area are defined by x and y coordinates (Figures 3-12 and 3-13). A source could represent the mouth of a river or a point within a river. Fluid elements emerge with velocities X_{veloc} and Y_{veloc}. Fluid discharge rates determine the rate at which individual fluid elements emerge form source. Sources may be widely distributed over the grid to represent separate streams, or grouped together to represent a single larger river.

NUMBER OF SOURCES can range from one to 20 sources. Each may represent a single fluvial source (Figure 7-5), or may be closely grouped to collectively represent larger rivers.

X and **Y** are x-y coordinates in meters that defines location of fluid sources the grid, with the origin in the lower left corner of the grid (Figure 3-12).

Xveloc and **Yveloc** are vector components describing the magnitude and direction of flow of fluid elements emerging from sources (Figure 7-2), in meters per second.

Discharge in m^3/sec defines rate at which fluid emerges from a fluid source. SEDSIM presently assumes that discharge rates at sources do not change during an

215

experiment. Discharge is the product of fluid velocity multiplied by cell width and flow depth, and should be commensurate with the volume of the river being simulated. Discharge rates of four rivers that form major deltas are listed in Table D-4.

Table D-4 Average fluid and sediment discharge rates for four rivers that form major deltas compiled from Smith (1966), Wright and Coleman (1973), and Coleman and Wright (1975).

DELTA NAMES:	Mississippi	Danube	Niger	Nile
Country	USA	Romania	Nigeria	Egypt
Number of river mouths	22	14	11	2
Fluid discharge (m^3/sec)	15,631	6,250	8,769	2,250
Sediment concentration (kg/m^3)	1.049	0.507	0.092	0.846

SEDIMENT LOAD

Sediment load is sediment carried by fluid elements as they emerge from a source (Figure 7-4) and is defined as a mixture of four grain types, with the total sediment load given as a concentration in kg/m^3. Concentrations in rivers of four major deltas are also listed in Table D-4.

% COARSE, % MEDIUM, % FINE, and **% FINEST** define percentages of the four grain types contributing to total load of sediment in fluid element at source locations.

CONCENTRATION in kg/m^3 is the sediment concentration in fluid elements as they emerge from fluid sources, with examples from typical deltas shown in Table D-4.

TOPOGRAPHY

Topography at the outset of an experiment is given by an array employing an x, y, z coordinate system (Figures 3-12 and 7-7), with z values representing elevations relative to sea level. A partial listing of grid-node elevations are shown in Table D-1. Subaerial elevations are positive and submerged elevations are negative. Users can generate simple topographic surfaces or provide elevations digitized from maps or digital terrain files.

GRID SIDE in meters establishes length Δx and Δy of cells in the topographic grid, thereby establishing the scale of the simulation grid (Figure 3-12).

NROW is the number of rows of grid cells along the x axis (M in Figure 3-12)

NCOL is the number of columns of grid cells along the y axis.(N in Figure 3-12)

Mathematical symbols used in text

Quantity	Symbol	Units
Angle of friction, or angle of maximum repose ($\tan\phi$)	ϕ	degrees
Angle of incidence of approaching waves	α	degrees
Angle of incidence of breaking waves	α_b	degrees
Angular direction of sediment transport	α	degrees
Area of grid cell	$\Delta x \Delta y$	m^2
Celerity or phase velocity	C	m/s
Coefficient of wave friction (Jonnson ,1966)	f_w	dimensionless
Coefficients of sediment transport in wave-power equations	$K, K1, K2, K3$	dimensionless
Coefficient of calibration for transport efficiencies	C_f	s^3/kg
Coefficient of drag	C_D	dimensionless
Coordinates, spatial	x, y, z	m
Density	ρ	kg/m^3
Density of sediment	ρ_s	kg/m^3
Density of fluid	ρ_f	kg/m^3
Density of water	ρ_w	kg/m^3
Depth water, including surface displacement by waves	D	m
Depth of water from mean sea level	d	m
Depth of water at breaker zone	d_b	m
Depth of moving bedload or "depth of disturbance"	d_m	m

217

Mathematical symbols used in text

Diameter (average) of sediment in bedload	D	m
Diameter (average) of sea bed	K	m
Diameter of long axis of orbital near bottom	do_{max}	m
Displacement, or elevation of water surface	η	m
Efficiency of bed load transport	ε_b	dimensionless
Efficiency of suspended load transport	ε_s	dimensionless
Energy	$\mathbf{E}$	kg m²/s²
Energy per unit length wave crest, per unit length of beach	E	kg/s²
Energy density per unit area of wave crest, at breaker zone	E_b	kg/s²
Energy density per unit area of wave crest, in deep water	E_o	kg/s²
Fall velocity	W	m/s
Grain types (maximum of four used here)	ks	dimensionless
Gravity	g	m/s²
Group velocity at breaker zone	C_b	m/s
Group velocity or velocity of energy propagation	C_g	m/s
Indices for x, y, z spatial directions	i, j, k	dimensionless
Momentum	$\mathbf{M}$	kg m/s
Momentum flux along direction of wave advance, parallel to x-axis	S_{xx}	kg/s²
Momentum flux along direction of wave crests, parallel to y-axis	S_{yy}	kg/s²
Number of columns in grid	N	dimensionless
Number of rows in grid	M	dimensionless
Orbital velocity under wave peak	U_{peak}	m/s
Orbital velocity under wave trough	U_{trough}	m/s
Obital velocity, maximum	U_{max}	m/s
Orbital velocity near sea bottom, instantaneous during passing of wave having period T	u_t	m/s
Phase angle	θ	radians
Phase velocity, or celerity, in deep water	C_d	m/s
Phase velocity, or celerity, in shallow water	C_s	m/s
Pivot angle	Φ	degrees
Power	$\mathbf{P}$	kg m²/s³
Power or wave energy flux in longshore direction, per unit length of shore	P_l	kg m/s³
Pressure	P	kg /(m s²)
Pressure at sea bottom	P_{-d}	kg /(m s²)
Pressure, dynamic wave-induced pressure at bottom	P_{dyn}	kg /(m s²)
Reynolds number	R_e	dimensionless
Sediment transport rate	Q	m³/s
Sediment transport rate, instantaneous during passing of wave having period T	$Q_{(t)}$	m³/s

Sediment transport rate, maximum in grid area	Q_{max}	m³/s
Sediment transport rate over total width of surf zone	Q_{tot}	m³/s
Sediment transport direction	α	degrees
Shear stress	τ	kg /(m s²)
Shear stress, critical threshold stress for motion	τ_c	kg /(m s²)
Shear stress at water surface	τ_s	kg /(m s²)
Shear stress at bottom surface	τ_b	kg /(m s²)
Shear stress, instantaneous, exerted near sea bottom by steady and oscillatory currents	$\tau_{o(t)}$	kg /(m s²)
Shear stress components, including bottom shear and surface shear	τ_{xx} , τ_{yx} , τ_{zx}	kg /(m s²)
Shear velocity	$U*$	m/s
Shields' parameter	Ψ	dimensionless
Shields parameter, instantaneous during passing of wave having period T	$\Psi_{(t)}$	dimensionless
Slope of beach	β	degrees
Time	t	s
Time step	Δt	s
Topographic elevation	z	m
Transport efficiencies of four grain types	$\varepsilon_{1,2,3,4}$	dimensionless
Transport efficiency of grain type ks	ε_{ks}	dimensionless
Velocities of fluid in x , y, and z directions	U, V, W	m/s
Velocity components in the x, y, and z directions -at water surface (free-surface boundary)	u_η , v_η , w_η	m/s
Velocity components in the x, y, z directions - at sea bottom (rigid surface boundary)	u_{-d} , v_{-d} , w_{-d}	m/s
Velocity of depth-averaged mean currents	U_{mean}	m/s
Velocity of grains moving as bedload and suspended load	V_g	m/s
Velocity, average longshore within surf zone	v_l	m/s
Velocity of near-bottom flow for calculating shear stress	υ	m/s
Viscosity of fluid	μ	kg /(m s)
Viscosity, kinematic	v	m²/s
Volume of sediment moving to or from a cell	Vol	m³
Volume of grain type ks moving to or from a cell	Vol_{ks}	m³
Volume of moving grain type ks parallel to y axis at cell location i, j	$QL_{ksi,j}$	m³
Volume of moving grain type ks, parallel to x axis at cell location i, j	$QC_{ksi,j}$	m³
Wave amplitude	A	m
Wave angle in deep water	β_d	degrees
Wave angle in shallow water	β_s	degrees
Wave frequency	ω	radians/s

Mathematical symbols used in text

Wave frequency (time averaged)	ϖ	radians/s
Wave height	H	m
Wave height in deep water	H_o	m
Wave height at breaker zone	H_b	m
Wave height in shallow water	H_s	m
Wavelength	L	m
Wavelength of breaking waves	L_b	m
Wavelength in deep water	L_o	m
Wave number	k	dimensionless
Wave period	T	s
Width of surf zone between breaker zone & shore	X_s	m
Wind speed	w	m/s
Wind speed required to initiate waves	w_c	m/s

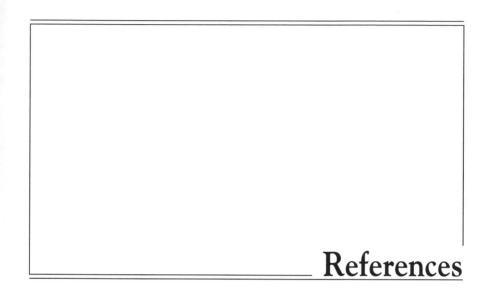

References

Ackers, P. and White, W. R., 1973, Sediment transport: a new approach and analysis, *Proc. Hydraulics Division*, Am. Soc. Civil Engrs., WY 11, p. 2041-2060.

Airy, G. B., 1845, Tides and waves, *Encyclopedia Metropolitana V*, Article 192, p. 241-396.

Allen, J. R., 1985, Field evaluation of beach profile response to wave steepness as predicted by the Dean model, *Coastal Eng.*, v. 9, p. 71-80.

Arthur, R. S., 1950, Refraction of shallow water waves: the combined effect of currents and underwater topography, *Trans. Am. Geophys. Union*, v. 31, no. 4, p. 549-552.

Arthur, R. S., 1962, A note on the dynamics of rip currents, *J. Geophysical Research*, v. 67, p. 2777-2779.

Bagnold, R. A., 1940, Beach formation by waves; some model experiments in a wave tank, *in:* J. S. Fisher and R. Dolan (eds.), *Beach Processes and Coastal Hydrodynamics*, Benchmark Papers in Geology, Dowden, Hutchinson & Ross, Stroudsburg, Pennsylvania, v. 39, p. 281 - 303.

Bagnold, R. A., 1956, The flow of cohesionless grains in fluids, *Phil. Trans. Royal Soc. London*, Ser. A, no. 249, p. 235-297.

Bagnold, R. A., 1962, Autosuspension of transported sediment; turbidity currents, *Proc. Royal Soc. of London*, Series A, v. 265, no. 1322, p. 315- 319.

References

Bagnold, R. A., 1963, Mechanics of marine sedimentation, in: M. N. Hill (ed.), *The Sea: Ideas and Observations on Progress in the Study of the Seas*, John Wiley & Sons, Interscience, New York, v. 3, p. 507- 528.

Bagnold, R. A., 1966, An approach to the sediment transport problem from general physics, *U. S. Geological Survey Professional Paper*, no. 422-I, 37 p.

Bailard, J. A., 1981, An energetics total load sediment transport model for a plane sloping beach, *J. Geophys. Research*, v. 86, p. 10938-10954.

Bailard, J. A., 1984, A simplified model for longshore sediment transport, *Proc. 19th Int. Conf. Coastal Eng.*, Am. Soc. Civil Engrs., New York, v. 2, p. 1454-1469.

Bakker, W. T. and Th. van Doorn, 1978, Near-bottom velocities in waves with a current, *Proc. 16th Int. Conf. Coastal Eng.*, Am. Soc. Civil Engrs., New York, v. 2, p. 1394-1413.

Bascom, W. N., 1951. The relationship between sand size and beach-face slope, *Trans. Am. Geophys. Union*, v. 32, no. 6, p. 866-874.

Bascom, W. N., 1954, Characteristics of natural beaches, *Proc. 4th Int. Conf. Coastal Eng.*, Am. Soc. Civil Engrs., New York, v. 1 p. 163 - 180.

Bascom, W. N., 1959, The relationship between sand size and beach face slope, in: J. S. Fisher and R. Dolan (eds.), *Beach Processes and Coastal Hydrodynamics*, Benchmark Papers in Geology, Dowden, Hutchinson & Ross, Stroudsburg, Pennsylvania, v. 39, p. 313 - 315.

Bates, C. C., 1953, Rational theory of delta formation, *Am. Assoc. Petroleum Geol. Bull.*, v. 37, no. 9, p. 2119-2162.

Bijker, E. W., 1971, Longshore transport computations, *Proc. Am. Soc. Civil Engrs., Journal of the Waterways, Harbours, and Coastal Eng. Division*, no. 97 (WW4), p. 687-701.

Bijker, E. W., E. van Hijum, and P. Vellinga, 1976, Sand transport by waves, *Proc. 15th Int. Conf. Coastal Eng.*, Am. Soc. Civil Engrs., New York, v. 2, p. 1149-1163.

Birkemeier, W. A., 1984, Time scales of nearshore bar changes, *Proc. 19th Int. Conf. Coastal Eng.*, Am. Soc. Civil Engrs., New York, v. 2, p. 1507-1520.

Birkemeier, W. A. and R. A. Dalrymple, 1975, Nearshore water circulation induced by wind and waves, *Proc. of the Symposium on Modeling Techniques*, Am. Soc. Civil Engrs., San Francisco, p. 1062-1081.

Boer, S., H. J. Vriend, and H. G. Wind, 1984, A system of mathematical models for the simulation of morphological features in the coastal zone, *Proc. 19th Int. Conf. Coastal Eng.*, Am. Soc. Civil Engrs., New York, v. 2, p. 1437-1453.

Bonham-Carter, G., and A. J. Sutherland, 1968, Mathematical Model and FORTRAN IV program for computer simulation of deltaic sedimentation, *Computer Contr. 24*, Kansas Geological Survey, Lawrence, Kansas, 56 p.

Bowen, A. J., 1969, The generation of longshore currents on a plane beach, *J. Marine Res.*, v. 27, p. 206-215.

Bowen, A. J., 1980, Simple models of nearshore sedimentation; beach profiles and longshore bars, *in:* S. B. McCann (ed.), *The Coastline of Canada*, Geol. *Survey Canada Paper No. 80 10*, Canadian Govt. Pub. Centre, Hull, Quebec, p. 1-11.

Bowen, A. J., D. L. Inman, and V. P. Simmons, 1968, Wave setdown and setup, *J. Geophys. Res.*, v. 73, no. 8, p. 2569-2577.

Bowen, A. J., and J. C. Doering, 1984, Nearshore sediment transport: estimates from detailed measurements of the nearshore velocity field, *Proc. 19th Int. Conf. Coastal Eng.*, Am. Soc. Civil Engrs., New York, v. 2, p. 1703-1713.

Boyd, D. R., and B. F. Dyer, 1964, Frio barrier bar system of South Texas, *Gulf Coast Assoc. Geol. Socs. Trans.*, v. 14, p. 309-322.

Brenninkmeyer, B. M., 1975, Frequency of sand movement in the surf zone, *Proc. 14th Int. Conf. Coastal Eng.*, Am. Soc. Civil Engrs., New York, v. 2, p. 812-827.

Bridge, J. S., and D. F. Dominic, 1984, Bed load grain velocities and sediment transport rates, *Trans Am. Geophys. Union*, p. 476-490.

Bruno, R. O., and C. G. Gable, 1976, Longshore transport at a total littoral barrier, *Proc. 15th Int. Conf. Coastal Eng.*, Am. Soc. Civil Engrs., New York, v. 2, p. 1203-1223.

Bruno, R. O., R. G. Dean, C. G. Gable and T. L. Walton, 1981, Longshore sand transport study at Channel Island harbour, California, *U.S. Army Coastal Engineering Research Center (CERC) Technical Paper 81-2*, 48 p.

Burrows, R., and T. S. Hedges, 1985, The influence of currents on ocean wave climates, *Coastal Eng.*, v. 9, p. 247-260.

Caldwell, J. M., 1956, Wave action and sand movement near Anaheim Bay, California, *U. S. Army Beach Erosion Board, Tech. Mem.*, 68, 21 p.

Carmel, Z., D. L. Inman and A. Golik, 1984, Transport of Nile sand along the southeastern Mediterranean coast, *Proc. 19th Int. Conf. Coastal Eng.*, Am. Soc. Civil Engrs., New York, v. 2, p. 1282-1290.

Chandramohan, P., B. U. Nayak, and V. S. Raju, 1988, Application of longshore transport equations to the Andrha Coast, East Coast of India, *Coastal Eng.*, v. 12, p. 285-297.

Chapman, D. M., 1981, Coastal erosion and the sediment budget, with special reference to the Gold Coast, Australia, *Coastal Eng.*, v. 4, p. 207-227.

Clifton, H. E., 1976, Wave-generated structures- a conceptual model, *in:* R. A. Davis and R. L. Ethington, (eds.), *Beach and Nearshore Sedimentation*, Soc. Econ. Paleontol. Mineral., Spec. Publ. 24 , p. 126-148.

Clifton, H. E., R. E. Hunter, and R. L. Phillips, 1971, Depositional structures and processes in the non-barred high-energy nearshore, *J. Sedimentary Petrol.*, v. 41, no. 3, p. 651 - 670.

Clifton H. E., and J. R. Dingler, 1984, Wave-formed structures and paleo-environmental reconstruction, *in:* B. Greenwood and R. A. Davis Jr. (eds.), *Hydrodynamics and Sedimentation in Wave-dominated Coastal Environments- Developments in Sedimentology*, Elsevier, New York, v. 39, p. 165-196.

Coakley, J., and M. G. Skafel, 1982, Suspended sediment discharge on a non-tidal coast, *Proc. 18th Int. Conf. Coastal Eng.*, Am. Soc. Civil Engrs., New York, v. 2, p. 1288-1304.

Colby, B. R., 1964, Practical computations of bed-material discharge, *J. Hydraulics Div.*, Am. Soc. Civil Engrs., v. 90, no. HY2, p. 217-246.

Coleman, J. M., and S. M. Gagliano, 1964, Cyclic sedimentation in the Mississippi River deltaic plain, *Gulf Coast Assoc. Geol. Socs. Trans.*, v. 14, p. 67-80.

Coleman, J. M., and L. D. Wright, 1975, Modern river deltas: variability of processes and sand bodies, *in:* M. L. Broussard (ed.), *Deltas- Models for Exploration*, Houston Geological Society, Houston, p. 99-149.

Cook, D. O., and D. S. Gorsline, 1972, Field observations of sand transport by shoaling waves, *Marine Geol.*, v. 13, p. 31-55.

Copeland, G. J. M., 1985, Practical radiation stress calculations connected with equations of wave propagation, *Coastal Eng.*, v. 9, p. 195-219.

Cornaglia, P., 1898, *Sul Regime delle Spaggie e sulla Regolazione dei Porti*. Torino, Summarized in Munch-Petersen (1950), Munch-Peterson's littoral drift formula, *U. S. Army Beach Erosion Board, Bull.*, no. 4, p.1-31.

Davies, A. M., 1987, On extracting current profiles from vertically integrated numerical models, *Coastal Eng.*, v. 11, p. 445-477.

Dean, R. G., E. Berek, C. G. Gable, and R. J. Seymour, 1982, Longshore transport rate determined by an efficient trap, *Proc. 18th Int. Conf. Coastal Eng.*, Am. Soc. Civil Engrs., New York, v. 2, p. 954-968.

Dean, R. G., and M. Perlin, 1986, Intercomparison of nearbottom kinematics by several wave theories and field and laboratory data, *Coastal Eng.*, v. 9, p. 399-437.

Dean, R. G., E. Berek, K. R. Bodge, and C. G. Gable, 1987, NSTS measurements of total longshore transport, *in: Coastal Sediments 1987*, Am. Soc. Civil Engrs., New York, p. 652-667.

DeVriend, H. J. D., 1987, 2DH Mathematical modeling of morphological evolutions in shallow water, *Coastal Eng.*, v. 11, p. 1-27.

DeVriend, H. J. D., and M. J. F. Stive, 1987, Quasi-3D modeling of nearshore currents, *Coastal Eng.*, v. 11, p. 565-601.

Dingler, J. R., 1979, The threshold of grain motion under oscillatory flow in a laboratory wave channel, *J. Sedimentary Petrol.*, v. 49, no.1, p. 287-294.

Dingler, J. R., 1986, Personal Communication, U. S. Geological Survey, Menlo Park, California.

Dingler, J. R., and D. L. Inman, 1976, Wave-formed ripples in nearshore sands, *Proc. 15th Int. Conf. Coastal Eng.*, Am. Soc. Civil Engrs., New York, v. 3, p. 2109-2126.

Dobson, R. S., 1967, Some applications of digital computers to hydraulic engineering problems, *Dept. Civil Engineering, Stanford Univ. Tech. Report TR-80*, Stanford, California, 172 p.

Douglass, S. L., and J. R. Weggel, 1988, Laboratory experiments of the influence of wind on nearshore breaking, *Proc. 21st Int. Conf. Coastal Eng.*, Am. Soc. Civil Engrs., New York, v. 1, p. 632-643.

Duane, D. B. and W. R. James, 1980, Littoral transport in the surf zone elucidated by a Eulerian tracer experiment, *J. Sedimentary Petrol.*, v. 50, p. 929-942.

Du Boys, M. P., 1879, Etudes du regime et laction exercee par les eaux sur un lit a fond de graviers indefiniment affouliable, *Annales des Ponts et Chausses*, Ser. 5, v. 18, p. 141-195.

Eagleson, P. S., and R. G. Dean, 1961, Wave-induced motion of bottom sediment particles, *Am. Soc. of Civil Engs. Trans.*, v. 126, p. 1162-1189.

Ebersole, B. A., and R. A. Dalrymple, 1979, A numerical model for nearshore circulation including convective accelerations and lateral mixing, *Dept. Civil Engineering Tech. Report no. 4*, Univ. of Delaware, Newark, Delaware, *Ocean Engineering Report 21*, Office of Naval Research Geography Programs, 87 p.

Einstein, H. A., 1948, Movement of beach sand by water waves, *Trans. Am. Geophys. Union*, v. 29, p. 653-655.

Einstein, H. A., 1950, The bed load function for sediment transportation in open channel flows, *U. S. Dept. of Agric. Tech. Bull.*, v. 1026., 29 p.

Einstein, H. A., 1972, A basic description of sediment transport on beaches, *in:* R. E. Meyer (ed.), *Waves on Beaches and Resulting Sediment Transport*, Academic Press, New York, p. 53-92.

Felder, W. N., and J.S. Fisher, 1980, Simulation model of analysis of seasonal beach cycles, *Coastal Eng.*, v. 3, p. 269-282.

Fisher, W. L., 1969, Facies characterization of Gulf Coast Basin systems, with some Holocene analogues, *Gulf Coast Assoc. Geol. Socs. Trans.*, v. 14, p. 239 - 261.

References

Fisher, W. L., L. F. Brown, Jr., A. J. Scott, and J. H. McGowen, 1969, *Delta systems in the exploration for oil and gas, a research colloquium*, Bureau of Economic Geology, Univ. Texas, Austin, 212 p.

Fleming, C. A., and J. N. Hunt, 1970, Application of sediment transport models, *Proc. 12th Int. Conf. Coastal Eng.*, Am. Soc. Civil Engrs., New York, v. 2, p. 1184-1203.

Forsythe, G. E., M. A. Malcolm, and C. B. Moler, 1977, *Computer Methods for Mathematical Computations*, Prentice-Hall, Englewood Cliffs, New Jersey, p. 156-166.

Fox, W. T., 1985, Modeling coastal environments, *in:* R. A. Davis Jr. (ed.), *Coastal Sedimentary Environments*, Springer Verlag, New York, p. 665-705.

Fox, W. T., and R. A. Davis Jr., 1979, Computer models of wind, waves, and longshore currents during a coastal storm, *Math. Geol.*, v. 11, no. 2, p. 143-163.

Frazier, D. E., 1967, Recent deltaic deposits of the Mississippi River: their development and chronology, *Gulf Coast Assoc. Geol. Socs. Trans.*, v. 17, p. 287-311.

Galloway, W. E., 1975, Process framework for describing the morphologic and stratigraphic evolution of deltaic depositional systems, *in:* M. L. Broussard (ed.), *Deltas- Models for Exploration*, Houston Geol. Soc., Houston, p. 87-98.

Galloway, W. E., and E. S, Cheng, 1985. Reservoir facies architecture in a microtidal barrier system-Frio Formation, Texas Gulf Coast, *Bureau of Economic Geology Report of Investigations No. 144*, Univ. Texas, Austin, 36 p.

Galvin, C. J., and P. Vitale, 1976, Longshore transport prediction- SPM 1973 Equation, *Proc. 15th Int. Conf. Coastal Eng.*, Am. Soc. Civil Engrs., New York, v. 2, p. 1133-1149.

Gordon, A. D., and J. G. Hoffman, 1984, Sediment transport on the south-east Australian continental shelf, *Proc. 19th Int. Conf. Coastal Eng.*, Am. Soc. Civil Engrs., New York, v. 2, p. 1952-1967.

Greenwood, B., and R. A. Davis Jr. (eds.), 1984, Hydrodynamics and Sedimentation in Wave-dominated Coastal Environments, *Developments in Sedimentology, v. 39*, Elsevier, New York, 263 p.

Greer, M. N. and Madsen, O. S., 1978, Longshore sediment transport data: A Review, *Proc. 16th Int. Conf. Coastal Eng.*, Am. Soc. Civil Engrs., New York, v. 2, p. 1563-1576.

Gould, H. R., and E. McFarlan, Jr., 1984, Geologic history of the Chenier plain, Southwestern Louisiana, *in:* E.C. Roy Jr. (ed.), *Readings in Gulf Coast Geology*, Gulf Coast Assoc. Geol. Socs., v. 5, Plate 2, p. 43.

Handin, J. W., and J. C. Ludwick, 1950, Accretion of beach sand behind a detached a breakwater, *U. S. Army Beach Erosion Board, Tech. Mem., 16*, 13 p.

Harbaugh, J. W., and G. Bonham-Carter, 1970, *Computer Simulation in Geology*, Wiley-Interscience, New York, 575 p.

Hardy, T. A., and N. C. Kraus, 1988, Coupling stokes and cnoidal wave theories in a nonlinear refraction model, *Proc. 21st Int. Conf. Coastal Eng.*, Am. Soc. Civil Engrs., New York, v. 2, p. 588-601.

Harlow, F. H., 1964, The particle-in-cell computing method for fluid dynamics, *in*: B. Alder (ed.), *Comput. Physics*, v. 3. Academic Press, New York, p. 319-343.

Hattori, M., 1982, Field study on onshore-offshore transport, *Proc. 18th Int. Conf. Coastal Eng.*, Am. Soc. Civil Engrs., New York, v. 2, p. 923-940.

Holman, R. A., and A. J. Bowen, 1982, Bars, bumps, and holes: models for the generation of complex beach topography, *J. Geophys. Res.*, v. 87, no. C1, p. 457-468.

Holman, R. A., and A. H. Sallenger Jr. 1986. High-energy nearshore processes, *EOS*, v. 67, no. 49, pp. 1369-1371.

Holman, R. A., 1991, Personal communication, Professor of Oceanography, Oregon State University, Corvallis, Oregon.

Horikawa, K., 1988, *Nearshore Dynamics and Coastal Processes: Theory, Measurement, and Predictive Models*, Univ. Tokyo Press, Tokyo, 522 p.

Horikawa, K., and T. Sasaki, 1972, Field observations of nearshore current systems, *Proc. 13th Int. Conf. Coastal Eng.*, Am. Soc. Civil Engrs., New York, v. 1, p. 635-652.

Hunter, R. E., H. E. Clifton, and R. L. Phillips, 1979, Depositional processes, sedimentary strcutures, and predicted vertical sequences in barred nearshore systems, southern Oregon coast, *J. Sedimentary Petrol.*, v. 49, no. 3, p. 711-726.

Ingle, J. C., Jr., 1966, *The Movement of Beach Sand*, Elsevier, New York, 221 p.

Ingle, J. C., Jr., 1986, Personal communication, Professor of Applied Earth Sciences, Stanford University, Stanford, California.

Inman, D. L., 1949, Sorting of sediments in the light of fluid mechanics, *J. Sedimentary Petrol.*, v. 19, no. 2, p. 51-70.

Inman, D. L., 1950, Report on beach study in the vicinity of Mugu Lagoon, California, *U. S. Army Beach Erosion Board, Tech. Mem.*, 14, 23 p.

Inman, D. L., and R. A. Bagnold, 1963, Littoral processeses, *in*: M. N Hill (ed.), *The Sea: Ideas and Observations on Progress in the Study of the Seas*, Wiley & Sons, New York, v. 3, p. 529-553.

Inman, D. L., and N. Nasu, 1956, Orbital velocity associated with wave action near the breaker zone, *U. S. Army Beach Erosion Board, Tech. Mem.*, 79, 43 p.

227

References

Inman, D. L., J. A. Zampol, T.E. White, D. M. Hanes, B. W. Waldorf, and K. A. Karstens, 1980, Field measurements of sand motion in the surf zone, *Proc. 17th Int. Conf. Coastal Eng.*, Am. Soc. Civil Engrs., New York, v. 2, p. 1215-1234.

Ippen, A. T., and P. S. Eagleson, 1955, A study of sediment sorting by waves shoaling on a plane beach, *U. S. Army Beach Erosion Board Tech. Mem. 63*, 83 p.

Ito, Y., and T. Katsutoshi, 1972, A method of numerical analysis of wave propagation-application to wave diffraction and refraction, *Proc. 13th Int. Conf. Coastal Eng.*, Am. Soc. Civil Engrs., New York, v. 1, p. 503-522.

Johnson, J. W., 1956, Dynamics of nearshore sediment movement, *Am. Assoc. Petrol. Geol. Bull.*, v. 40, no. 9, p. 221 - 232.

Jonsson, I. G., 1966, Wave boundary layers and friction factors, *Proc. 10th Int. Conf. Coastal Eng.*, Am. Soc. Civil Engrs., New York, v. 1, p. 127-148.

Kamphuis, J. W., and O. J. Sayao, 1982, Model tests on littoral sand transport rate, *Proc. 18th Int. Conf. Coastal Eng.*, Am. Soc. Civil Engrs., New York, v. 2, p. 1305- 1325.

Kamphuis, J. W., M. H. Davies, R. B. Nairn, and O. J. Sayao, 1986, Calculation of littoral sand transport rate, *Coastal Eng.*, v. 10, p. 1-21.

Kim, T. I., R. T. Hudspeth, and W. Sulisz, 1986, Circulation kinematics in nonlinear laboratory waves, *Proc. 20th Int. Conf. Coastal Eng.*, Am. Soc. Civil Engrs., New York, v. 1, p. 381-395.

King, C. A. M., 1951, Depth of disturbance of sand on sea beaches by waves, *J. Sedimentary Petrol.*, v. 21, no. 3, pp 131-140.

King, D. B., Jr., and R. J. Seymour, 1989, State of the art in oscillatory sediment transport models, *in:* R. J. Seymour (ed.), *Nearshore Sediment Transport*, Plenum Press, New York, p. 371-385.

Knoth, J. S., and D. Nummedal, 1977, Longshore sediment transport using fluorescent tracers, *Coastal Sediments '77*, Am. Soc. Civil Engrs., New York, p. 383-398.

Kolb, C. R., and J. R. van Lopkin, 1966, Depositional environments of the Mississippi River deltaic plain-southeastern Louisiana, *in:* M. L. Shirley and J. A. Ragsdale (eds.), *Deltas in Their Geologic Framework*, Houston Geol. Soc., Houston, Tex., p. 17-61.

Koltermann, C. E. and S. M. Gorelick, 1990a, Geologic modelling of heterogeneity in alluvial fan deposits influenced by paleoclimate variability, *EOS*, v.71, no. 17 p. 508.

Koltermann, C. E. and S. M. Gorelick, 1990b, Geologic modelling of spatial variability in sedimentary environments: stochastic models of groundwater transport, *28th International Geological Congress*, v. 2, p. 2-208. Washington D.C.

Komar, P. D., 1969, *The Longshore Transport of Sand on Beaches* [unpublished PhD thesis], Dept. Oceanography, University of California, San Diego, 143 p.

Komar, P. D., 1971, The mechanics of sand transport on beaches, *J. Geoph. Res.*, v. 76, no. 3, p. 713- 721.

Komar, P. D., 1973, Computer models of delta growth due to sediment input from rivers and longshore tranport, *Geol. Soc. of Am. Bull.*, v. 84, p. 2217-2226.

Komar, P. D., 1976, *Beach Processes and Sedimentation*, Prentice- Hall, Englewood Cliffs, New Jersey, 429 p.

Komar, P. D., 1977, Mechanics of marine sedimentation, *in:* E.D. Goldberg and I. N. McCave (eds.), *The Sea: Ideas and Observations on Progress in the Study of the Seas- Marine Modelling*, Interscience, John Wiley & Sons, New York, v. 6, p. 499-513.

Komar, P. D., 1988, Environmental controls on littoral sand transport rate, *Proc. 21st Int. Conf. Coastal Eng.*, Am. Soc. Civil Engrs., New York, v. 2, p. 1238-1252.

Komar, P. D., 1989, Physical processes of waves and currents and the formation of marine placers, CRC *Critical Reviews in Aquatic Sciences*, Oregon Sea Grant, Oregon State University, v. 1, issue 3, p. 393-423.

Komar, P. D., 1991, Littoral sediment transport, *in:* J. B. Herbich (ed.), *Handbook of Coastal and Ocean Engineering*, Gulf Pub. Co., New York, p. 681-714.

Komar, P. D., 1991, Personal communication, Professor of Oceanography, Oregon State University, Corvallis, Oregon.

Komar, P. D., and D. L. Inman, 1970, Longshore sediment transport on beaches, *J. Geophys. Res.*, v. 75, p. 5914-5927.

Komar, P. D., and K. E. Clemens, 1986, The relationship between a grain's settling velocity and threshold of motion under unidirectional currents, *J. Sedimentary Petrol.*, v. 56, no. 2, p. 258-266.

Komar, P. D., and K. E. Clemens, 1988, Oregon beach-sand compositions produced by the mixing of sediments under a transgressive sea, *J. Sedimentary Petrol.*, v. 58, no. 3, p. 519- 529.

Komar, P. D., and M. C. Miller, 1973, The threshold of sediment movement under oscillatory water waves, *J. Sedimentary Petrol.*, v. 43, no. 4, p. 1101- 1110.

Kraus, N. C., K. J. Gingerich, and J. D. Rosati, 1988, Toward an empirical formulae for longshore sand transport, *Proc. 21st Int. Conf. Coastal Eng.*, Am. Soc. Civil Engrs., New York, v. 2, p. 1182-1197.

References

Kraus, N. C., M. Isobe, H. Igarashi, T. O. Sasaki, and K. Horikawa, 1982, Field experiments of longshore sand transport in the surfzone, *Proc. 18th Int. Conf. Coastal Eng.*, Am. Soc. Civil Engrs., New York, v. 2, p. 969-988.

Krumbein, W. C., 1944, Shore processes and beach characteristics, *in:* J. S. Fisher and R. Dolan (eds.), *Beach Processes and Coastal Hydrodynamics*, Benchmark Papers in Geology, Dowden, Hutchinson & Ross, Stroudsburg, Pennsylvania, v. 39, p. 71-113.

Krumbein, W. C., 1964, *A Geologic Process-Response Model for Analysis of Beach Phenomena*, Northwestern University Geology Department, Technical Report No. 8, pp. 1-15.

Kurihara, Y., 1965, On the use of implicit and iterative methods for time integration of the wave equation, *Monthly Weather Rev.*, v. 5, p. 33-46.

Lagaaij, R., and F. P. H. W. Kopstein, 1964, Typical features of a fluviomarine offlap sequence, *in:* L. M. van Stratten (ed.), *Deltaic and Shallow Marine Deposits*, Elsevier, New York, p. 216-226.

Lakhan, V. C., 1982, WAVES: A FORTRAN IV program on the stochastic simulation of waves in the coastal environment, *Computers and Geosciences*, v. 8, no. 1. p. 45-60.

Lakhan, V. C., 1989, Computer simulation of the characteristics of shoreward propagating deep and shallow water waves, *in:* V. C. Lakhan and A. S. Trenhaile (eds.), *Applications in Coastal Modeling*, Elsevier Oceanography Series, v. 49, Elsevier, New York, p. 107-158.

Larue, D. K., and P. A. Martinez, 1989, Use of bed-form climb models to analyze geometry and preservation potential of clastic facies and erosional surfaces, Am. Assn. Petroleum Geol. Bull., v. 73, no. 1, p. 40-53.

Lee, K. K., 1975, Longshore currents and sediment transport in west shore of Lake Michigan, *Water Resource Res.*, v. 11, p. 1029-1032.

Lee, Y. H. 1991, Three-dimensional modelling of Arkansas river sedimentation, Abstract in the 75th Am. Assn. Petroleum Geol. Annual Convention, Dallas, April 7-10, 1991.

Lee, Y. H. and J. W. Harbaugh, 1992, Stanford's SEDSIM project: Dynamic three-dimensional simulation of geologic processes that affect clastic sediments, *in:* R. Pflug and J. W. Harbaugh (eds.), *Three-Dimensional Computergraphics in Modeling Geologic Structures and Simulating Geologic Processes*, Lecture Notes in Earth Sciences #41, Springer-Verlag, Berlin, p. 113-127.

Lee-Young, J. S. and J. F. A. Sleath, 1988, Initial motion in combined wave and current flows, *Proc. 21st Int. Conf. Coastal Eng.*, Am. Soc. Civil Engrs., New York, v. 2, p. 1140-1150.

Le Mehaute, B., 1961, A theoretical study of waves breaking at an angle with a shore line, *J. Geophys. Res.*, v. 66, no. 2, p. 495-499.

Lenhoff, L., 1982, Incipient motion of particles under oscillatory flow, *Proc. 18th Int. Conf. Coastal Eng.*, Am. Soc. Civil Engrs., New York, v. 2, p. 1555-1568.

230

Leontev, I. O., 1985, Sediment transport and beach equilibrium profile, *Coastal Eng.*, v. 9, p. 277-291.

Liu, P. and C. C. Mei, 1974, *Effects of Breakwater on Nearshore Currents Due to Breaking Waves*, Dept. Civil Engineering Rep. no. 192., Massachusetts Inst. of Technology, Cambridge, Massachusetts, 113 p.

Longuet-Higgins, M. S., 1952, On the statistical distribution of heights of sea waves, *J. Mar. Res.*, v. 11, no. 13, p. 245-266.

Longuet-Higgins, M. S., 1970, Longshore currents generated by obliquely incident sea waves, 1 and 2, *J. Geoph. Res.*, v. 75, no. 33, p. 6778 - 6801.

Longuet-Higgins, M. S., and R. W. Stewart, 1962, Radiation stress and mass transport in gravity waves; with applications to surf beats, *J. Fluid Mech.*, v. 13, p. 481-504.

Madsen, O. S., 1989, Transport determination by tracers- Tracer theory, *in:* R. J. Seymour (ed.), *Nearshore Sediment Transport*, Plenum Press, New York, p. 103-114.

Madsen, O. S., and W. D. Grant, 1976, *Sediment transport in the coastal environment*, Report No. 209, Mass. Instit. of Tech., Cambridge, Mass., 120 pp.

Madsen, O. S., and W. D. Grant, 1977, Quantitative description of sediment transport by waves, *Proc. 12th Int. Conf. Coastal Eng.*, Am. Soc. Civil Engrs., New York, v. 2, p. 1093-1112.

Martinez, P. A., 1987a, WAVE: Program for simulating onshore-offshore sediment transport in two dimensions, *Computers and Geosciences*, v.13, no. 5, p. 513-532.

Martinez, P. A., 1987b, *Simulation of Sediment Transport and Deposition by Waves, for Simulation of Wave Versus Fluvial-Dominated Deltas* [unpublished Masters thesis], Dept. of Applied Earth Sciences, Stanford University, Stanford, California, 404 p.

Martinez, P. A., 1992a, Three-dimensional simulation of littoral transport, *in:* R. Pflug and J. W. Harbaugh (eds.), *Three-Dimensional Computergraphics in Modeling Geologic Structures and Simulating Geologic Processes*, Lecture Notes in Earth Sciences #41, Springer-Verlag, Berlin, p.129-141.

Martinez, P. A., 1992b, *Simulating Nearshore Processes* [unpublished Ph.D. dissertation], Department of Applied Earth Sciences, Stanford University, Stanford, California, 300 p.

Martinez, P. A., and J. W. Harbaugh, 1989, Computer simulation of wave and fluvial-dominated nearshore environments, in: V. C. Lakhan and A. S. Trenhaile (eds.), *Applications in Coastal Modeling*, Elsevier Oceanography Series, v. 49, Elsevier, New York, p. 297-337.

McCave, I. N., 1972, Patterns of fine sediment dispersal, in: D. J. P. Swift, D. B. Duane, and O. H. Pilkey (eds.), *Shelf Sediment Transport*, Dowden, Hutchinson and Ross, Stroudsburg, Pennsylvania, p. 225-248.

References

Meyer-Peter, E., and R. Muller 1948, Formulas for bedload transport, *Proc. Third meeting of the Inter. Assoc. of Hydrology Res.*, Stockholm, p. 39-64.

Miller, M. C., I. N. McCave, and P. D. Komar, 1977, Threshold of sediment motion under unidirectional currents, *Sedimentology*, v. 24, p 507-527.

Miller, R. L. and J. M. Zeigler, 1958, A model relating dynamics of sediment pattern in equilibrium in the region of shoaling waves, breaker zone, and foreshore, *J. Geol.*, vol. 66, no. 4, p. 417-441.

Miller, R. L. and J. M. Zeigler, 1964, A Study of sediment distribution in the zone of shoaling waves over complicated bottom topography, *in:* R. L. Miller (ed.). *Papers in Marine Geology, Shepard Commemorative Volume*, Macmillan, New York, p. 133-153.

Miller, R. L. and R. J. Byrne, 1966, The angle of repose of a single grain on a fixed rough bed, *Sedimentology*, v. 6, p. 303-314.

Moore, G. W., and J. Y. Cole, 1960, Coastal processes in the vicinity of Cape Thompson, Alaska, *in:* Kachdoorian and others (eds.), *Geologic Investigations in Support of Project Chariot in the Vicinity of Cape Thompson, Northwestern Alaska- Preliminary Report*, U.S. Geol. Survey Trace Elements Investigation Report 753, p. 41-55.

Morrison, J. R., and R. C. Crooke, 1953, The mechanics of deep water, shallow water, and breaking waves, *U. S. Army Beach Erosion Board Tech. Mem. 40*, 14 p.

Munch-Petersen, 1950, Munch-Peterson's littoral drift formula, *U. S. Army Beach Erosion Board Bull.*, no. 4, p 1-31.

Munk, W. H., 1949, The solitary wave theory and its application to surf problems, *Ann. N.Y Acad. Sci.*, v. 51, p. 376-424.

Murray, S. P., 1972, Observations on wind, tidal, and density-driven currents in the vicinity of the Mississippi River delta, in: D. J. P. Swift, D. B. Duane, and O. H. Pilkey (eds.), *Shelf Sediment Transport*, Dowden, Hutchinson and Ross, Stroudsburg, Pennsylvania, p. 127-142.

Niederoda, A. W., C. M. Ma, A. Mangarella, R. H. Cross, S. R. Huntsman, and D. O. Treadwell, 1982, Measured and computed coastal ocean bedload transport, *Proc. 18th Int. Conf. Coastal Eng.*, Am. Soc. Civil Engrs., New York, v. 2, p. 1353- 1368.

Noda, E. K., 1972, Rip currents, *Proc. 13th Int. Conf. Coastal Eng.*, Am. Soc. Civil Engrs., New York, v. 1, p. 653-668.

Noda, E. K., 1984, Depositional effects of an offshore breakwater due to onshore-offshore sediment movement, *Proc. 19th Int. Conf. Coastal Eng.*, Am. Soc. Civil Engrs., New York, v. 2, p. 2009-2025.

Noda, E. K., C. J. Sonu, V. C. Rupert, and J. I. Collins, 1974, *Nearshore Circulations Under Sea Breeze Conditions and Wave-Current Interactions in the Surf Zone*, technical Report No. TC-149-4, Tetra Tech Inc., Pasadena, California, 213 p.

Oomkens, E., 1967, Depositional sequences and sand distribution in a deltaic complex, *Geol. en Mijnbouw*, v. 46e, p. 265-278.

Ozasa, H., and A. H. Brampton, 1980, Mathematical modelling of beaches backed by seawalls, *Coastal Eng.*, v. 4, p. 47-63.

Penland, S., and R. Boyd, D. Nummedal, H. Roberts, 1981, Deltaic barrier development on the Louisiana coast, *Gulf Coast Assoc. Geol. Socs. Trans. Supplement*, v. 31, p. 471-476.

Pflug, R., Klein, H., Ramshorn, C., Genter, M., Stärk, A, 1992, 3-D visualization of geologic structures and processes, *in*: R. Pflug and J. W. Harbaugh (eds.), *Three-Dimensional Computergraphics in Modeling Geologic Structures and Simulating Geologic Processes*, Lecture Notes in Earth Sciences #41, Springer-Verlag, Berlin, p. 29-39.

Phillips, O. M., 1966, *The Dynamics of the Upper Ocean, 1st ed.*, Cambridge Univ. Press, Cambridge, 1966, p. 45-48.

Pond, S., and G. L. Pickard, 1983, *Introductory Dynamical Oceanography*, Pergamon Press, New York, p. 209 - 234.

Price, W. A., 1954, Dynamic environments: reconnaissance mapping, geologic and geomorphic, of continental shelf of Gulf of Mexico, *Gulf Coast Assoc. Geol. Socs. Trans.*, v. 4, p. 75-107.

Putnam, J. A., W. H. Munk, and M. A. Traylor, 1949, The prediction of longshore currents, *Trans. Am. Geophys. Union*, v. 30, p. 337-345.

Ramsden, J. D. and J. H. Nath, 1988, Kinematics and return flow in a closed wave flume, *Proc. 21st Int. Conf. Coastal Eng.*, Am. Soc. Civil Engrs., New York, v. 1, p. 448-462.

Ramshorn, C., R. Ottolini, and H. Klein, (in press), Interactive three-dimensional display of simulated sedimentary basins, *in*: *Proceedings of the Eurographics Workshop on Visualization in Scientific Computing in Clamart, France, April 1990*; Springer-Verlag, Heidelberg.

Richmond, B. M., and A. H. Sallenger, Jr., 1984, Cross-shore transport of bimodal sands, *Proc. 19th Int. Conf. Coastal Eng.*, Am. Soc. Civil Engrs., New York, v. 2, p. 1997-2008.

Saville, T., Jr., 1950, Model study of sand transport along an infinitely long straight beach, *Trans. Am. Geophys. Union*, v. 31, no.4, p. 555-565.

Saxena, R. S., 1976, *Modern Mississippi Delta-Depositional Environments and Processes: A Guidebook Prepared for the AAPG/SEPM Field Trip- Mississippi Delta Flight*, Am. Assoc. Petroleum Geol., New Orleans, 125p.

References

Scott, N. III, 1987, *Modern Versus Ancient Braided Stream Deposits: A Comparison Between Simulated Sedimentary Deposits and the Ivishak Formation of the Prudoe Bay Field, Alaska* [unpublished Master'd thesis], Department of Applied Earth Sciences, Stanford University, Stanford, California, 103p.

Scruton, P. C., 1956, Oceanography of Mississippi delta sedimentary environments, *Am. Assoc. Petroleum Geologists Bull.*, v. 40, no.12 , p. 2864-2952.

Seymour, R. J., (ed.), 1989, *Nearshore Sediment Transport*, Plenum Press, New York, p. 387-401.

Seymour, R. J., and D. Castel, 1989, Modeling cross-shore transport, *in:* R. J. Seymour (ed.), *Nearshore Sediment Transport.* Plenum Press, New York, p. 387-401, 418 p.

Shepard, F. P., 1950a, Beach cycles in Southern California, *U. S. Army Beach Erosion Board Tech. Mem. no. 20*, 26 p.

Shepard, F. P., 1950b, Longshore current observations in Southern California, *U. S. Army Beach Erosion Board, Tech. Mem., 13*, 54 p.

Shepard, F. P., and D. L. Inman, 1950, Nearshore water circulation related to bottom topography and wave refraction, *Trans. Am. Geophys. Union*, v. 31, p. 196-212.

Shepard, F. P., K. O. Emery, and E. C. La Fond, 1941, Rip currents: a process of geological importance, *J. Geol.*, v. 49, p. 337-369.

Shields, I. A., 1936, Anwendung der Ahnlichkeitsmechanik und der Turbulenzforchung auf die Geschiebebemegung, *Mitteilungen der Preussischen Veruchsanstalt fur Wasserbau und Schiffbau*, Heft 26, 20 p., Berlin. Available as translation by W. P. Ott and J. C. van Uchelen, S.C.C. Cooperative Laboratory, California Inst. Technology, Pasadena, Calif.ornia.

Silvester, R., 1984, Fluctuations in Littoral Drift, *Proc. 19th Int. Conf. Coastal Eng.*, Am. Soc. Civil Engrs., New York, v. 2, p. 1291.

Sleath, J. F. A., 1978, Measurement of bedload in oscillating flow, *J. Waterways, Ports, Coastal, and Ocean Division*, Am. Soc. Civil Engrs., New York, 104 (WW4), paper 13960, p. 291-307.

Sleath, J. F. A., 1984, *Sea Bed Mechanics*, John Wiley & Sons, New York, p. 123- 175, 258.

Slingerland, R. L. 1977, The effects of entrainment on the hydraulic equivalence relationships of light and heavy minerals in sand, *J. Sedimentary Petrol.*, v. 47, no. 2, p. 753-770.

Slingerland, R. L. and N. D. Smith, 1986, Occurrence and formation of water-laid placers, *Ann. Rev. Earth Planet. Sci.*, v. 14, p. 113-147.

Slingerland, R. L. 1991, Personal communication, Department of Geosciences, Penn State University, Pennsylvania.

Smith, A. E. Jr., 1966, Modern deltas- Comparison maps (Appendix), *in:* M. L. Shirley and J. A. Ragsdale (eds.) *Deltas in Their Geologic Framework,* Houston Geological Society, Houston, p. 233-251.

Steidtmann, J. R., 1982, Size-density soring of sand-size spheres during depositon from bedload transport and implications concerning hydraulic equivalence, *Sedimentology,* v. 29, p. 877-883.

Sternberg, R. W., N. C. Shi, and J. P. Downing, 1984, Field investigations of suspended sediment transport in the nearshore zone, *Proc. 19th Int. Conf. Coastal Eng.,* Am. Soc. Civil Engrs., New York, v. 2, p. 1782-1797.

Sternberg, R. W., N. C. Shi, and J. P. Downing, 1989, Suspended sediment measurements, *in:* R.J. Seymour (ed.), *Nearshore Sediment Transport,* Plenum Press, New York, p. 231-257.

Stive, M. J. F., and J. A. Battjes, 1984, A model for offshore sediment transport, *Proc. 19th Int. Conf. Coastal Eng.,* Am. Soc. Civil Engrs., New York, v. 2, p. 1420 - 1436.

Stokes, G. G., 1847, On the theory of oscillatory waves, *Trans. Cambridge Phil. Soc., Suppl. Sci. Papers,* v. 8, no. 1, p. 441.

Swain, A., 1989, Beach profile development, *in:* V. C. Lakhan and A. S. Trenhaile (eds.), *Applications in Coastal Modeling,* Elsevier Oceanography Series, v. 49, Elsevier, Amsterdam, p. 215-232.

Swart, D. H., 1974, A schematization of onshore-offshore transport, *Proc. 14th. Int. Conf. Coastal Eng.,* Am. Soc. Civil Engrs., New York, v. 1, p. 884-900.

Swart, D. H., 1976, Predictive equations regarding coastal transport, *Proc. 15th. Int. Conf. Coastal Eng.,* Am. Soc. Civil Engrs., New York, v. 2, p. 1113-1132.

Swart, D. H., and C. A. Fleming, 1980, Longshore water and sediment movement, *Proc. 17th Int. Conf. Coastal Eng.,* Am. Soc. Civil Engrs., New York, v. 2, p. 1275-1297.

Tetzlaff, D. M., 1987, *A Simulation Model of Clastic Sedimentary Processes* [unpublished Ph.D. dissertation], Department of Applied Earth Sciences, Stanford University, Stanford, California, 345 p.

Tetzlaff, D. M., and J. W. Harbaugh, 1989, *Simulating Clastic Sedimentation,* Van Nostrand Reinhold, New York, 202 p.

Thornton, E. B., and R. T. Guza, 1989, Nearshore circulation: conservation equations for unsteady flow, *in:* R. J. Seymour (ed.), *Nearshore Sediment Transport,* Plenum Press, New York, p. 183-203.

Trowbridge, A. C., 1930, Building of the Mississippi delta, *Am. Assoc. Petroleum Geol. Bull.,* v. 14, no. 7, p. 867-901.

References

Tyler, N., and W. A. Ambrose, 1985, Facies architecture and production characteristics of strandplain reservoirs in the Frio Formation, Texas, *Bureau of Economic Geology Report of Investigations No. 146*, Univ. Texas, Austin, 42 p.

U. S. Army Coastal Engineering Research Center (CERC), 1977, *Shore Protection Manual (3rd ed.)*, U.S. Government Printing Office, Washington, D.C., v. 1, p. 1:1 - 4:131 ; v. 3, p. C:41 - C:44.

Van de Graaff, J., and J. Van Overeem, 1979, Evolution of sediment transport formulae in coastal engineering practice, *Coastal Eng.*, v. 3, p. 1-32.

Van de Graaff, J., and W. M. K. Tilmans, 1980, Sand transport by waves, *Proc. 17th Int. Conf. Coastal Eng.*, Am. Soc. Civil Engrs., New York, v.2 p. 1140 - 1157.

Van Dorn, W. G., 1953, Wind stress on an artificial pond, *J. Marine Res.*, v. 12, 1953, p. 249-276.

Vemulakonda, S. R., J. R. Houston, and H. L. Butler, 1982, Modeling longshore currents for field situations, *Proc. 18th Int. Conf. Coastal Eng.*, Am. Soc. Civil Engrs., New York, v.2, p.1659- 1675.

Vincent, C. E., 1979, Longshore sand transport rates- a simple model for the east Anglian coastline, *Coastal Eng.*, v. 3, p. 113-136.

Watts, G. M., 1953, A study of sand movement at South Lake Worth Inlet, Florida, *U. S. Army Beach Erosion Board, Tech. Mem.*, *42*, 24 p.

Watts, G. M., and R. F. Dearduff, 1954, Laboratory study of the effect of varying wave periods on beach profiles, *U. S. Army Beach Erosion Board, Tech. Mem.*, 53, p.

Weggel, J. R., 1972, Maximum breaker height, *J. Waterways, Harbors, and Coastal Eng.*, Am. Soc. Civil Engs., New York, v. 108, no. WW4, p. 529 - 548.

Weise, B. R., 1980, Wave-dominated delta systems of the Upper Cretaceous San Miguel Formation, Maverick Basin, South Texas, *Bureau of Economic Geology Report of Inv. No. 107*, University Texas, Austin, 39 p.

Wendebourg, J., 1991, Three-dimensional modelling of sedimentation and subsurface fluid flow applied to the Woodbine Progradational System of Southeast Texas, Abstract in the 75th Am. Assn. Petroleum Geol. Annual Convention, Dallas, April 7-10 ,1991.

Wendebourg, J. and J. W. D. Ulmer, 1992, Modeling compaction and isostatic compensation in SEDSIM for basin analysis and subsurface fluid flow, *in:* R. Pflug and J. W. Harbaugh (eds.), *Three-Dimensional Computergraphics in Modeling Geologic Structures and Simulating Geologic Processes*, Lecture Notes in Earth Sciences #41, Springer-Verlag, Berlin, p. 143-153.

White, T. E., and D. L. Inman, 1989, Measuring longshore transport by tracers, *in:* R.J. Seymour (ed.), *Nearshore Sediment Transport*, Plenum Press, New York, p. 287-312.

236

Whitford, D. J., and E. B. Thornton, 1988, Longshore currents forcing at a barred beach, *Proc. 21st Int. Conf. Coastal Eng.*, Am. Soc. Civil Engrs., New York, v. 1, p. 77-90.

Wiegel, R. L., 1964, *Oceanographical Engineering*, Prentice-Hall, Englewood Cliffs, New Jersey, p. 1-29, 150-178, 195 - 245, 341-381.

Willis, D. H., 1980, Evaluation of sediment transport formulae in coastal engineering practice-discussion, *Coastal Eng.*, v. 4, p. 177-181.

Wilson, W. S., 1966, A method for calculating and plotting surface wave rays, *U. S. Army Beach Erosion Board Tech. Mem. 17*, 57 p.

Wright, L. D., 1977, Sediment transport and deposition at river mouths: a synthesis, *Geol. Soc. America Bull.*, v. 88, p. 857-868.

Wright, L. D., and J. M. Coleman, 1973, Variations in morphology of major river deltas as functions of ocean wave and river discharge regimes, *Am. Assoc. Petroleum Geol. Bull.*, v. 57, no. 2, p. 370-398.

Zampol, J. A., and D. L. Inman, 1989, Suspended sediment measurements- discrete measurements of suspended sediments, *in:* R.J. Seymour (ed.), *Nearshore Sediment Transport*, Plenum Press, New York, p. 259-372.

237

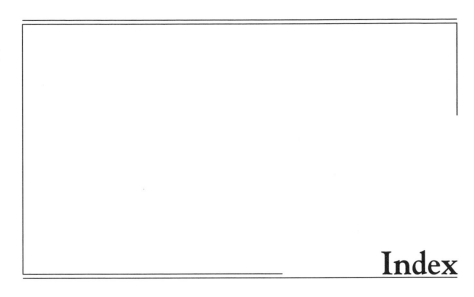

Index

G

L

M

V

W

X, Y, Z